Since taking his Ph.D
1966, FRITJOF CAPR
theoretical high-ener
Paris; the University c
ford University; and Imperial College in London. In addition to his technical research papers, he has written general articles about the relationship between modern physics and Eastern mysticism, and has lectured extensively about this topic to general student audiences in England and the United States. He is presently lecturing at the University of California, Berkeley.

"His synopsis of Eastern thought is noteworthy, and his presentation of modern physics for the layman among the clearest and most thorough I have seen."

—*Library Journal*

"Welcome, thought provoking, well-assembled, with lavish use of diagrams. It is a book that will grace the readers' shelves for years to come."

—*Times Educational Supplement*

"A clear and far-out discussion of such thorny concepts as space-time, quarks, and cosmic symmetries . . . Incredilbly lucid introductions to each spiritual discipline."

—*Village Voice*

"Will provide surprising insights to those readers who tend to scorn science. It also can speak to those scientists who choose to cast out mysticism."

—*East/West Journal*

"You may never read a more mind-expanding book than this . . . The equivalent of many university courses in philosophy, cosmology, astronomy and physics."

—*Spiritual Studies Center Booknews*

THE TAO OF PHYSICS

An Exploration of the Parallels
Between Modern Physics
and Eastern Mysticism

BY FRITJOF CAPRA

This low-priced Bantam Book
has been completely reset in a type face
designed for easy reading, and was printed
from new plates. It contains the complete
text of the original hard-cover edition.
NOT ONE WORD HAS BEEN OMITTED.

RL 11, IL 11+

THE TAO OF PHYSICS
A Bantam Book / published by arrangement with
Shambhala Publications, Inc.

PRINTING HISTORY
Shambhala Publications edition published January 1976
2nd printing . . . . June 1976 3rd printing . . . . . December 1976
4th printing . . . . . May 1977
Quality Book Club edition published July 1977
Bantam edition / November 1977
2nd printing . . . . November 1977 4th printing . . . . March 1979
3rd printing . . . . April 1978 5th printing . . . . April 1980

For information address: Shambhala Publications, Inc.,
1123 Spruce Street, P.O. Box 271, Boulder, Colorado 80302.

ISBN 0-553-14306-2

Published simultaneously in the United States and Canada

PRINTED IN THE UNITED STATES OF AMERICA

14 13 12 11 10 9 8 7 6 5

I dedicate this book to
Ali Akbar Khan
Carlos Castaneda
Geoffrey Chew
John Coltrane
Werner Heisenberg
Krishnamurti
Liu Hsiu Ch'i
Phiroz Mehta
Jerry Shesko
Bobby Smith
Maria Teuffenbach
Alan Watts
for helping me to find my path
and to Jacqueline
who has traveled with me
on this path most of the time.

ACKNOWLEDGMENTS

The author and publisher gratefully acknowledge permission to reproduce copyright illustrations on the following pages:

p. 1: Fermi National Laboratory, Batavia, Illinois;
pp. 40–41, 221, 222–223: CERN, Geneva, Switzerland;
p. 72: reprinted from *Zazen* by E. M. Hooykaas and B. Schierbeck, Omen Press, Tucson, Arizona;
p. 74: Estate of Eliot Elisofon;
p. 81: Gunvor Moitessier;
p. 82: reprinted from *The Evolution of the Buddha Image* by Benjamin Rowland Jr., The Asia Society, New York;
pp. 90, 174: Gulbenkian Museum of Oriental Art;
pp. 107, 246: reprinted from *Zen and Japanese Culture* by D. T. Suzuki, Bollingen Series LXIV, by permission of Princeton University Press;
p. 120: reprinted from *Physics in the Twentieth Century* by Victor Weisskopf, M.I.T. Press, Cambridge, Massachusetts;
pp. 187, 192, 210, 220, 225, 256: Lawrence Berkeley Laboratory, Berkeley, California;
pp. 217, 218: Argonne National Laboratory, Argonne, Illinois;
p. 231: reprinted from *The Arts of India* by Ajit Mookerjee, Thames and Hudson, London.

CONTENTS

It is probably true quite generally that in the history of human thinking the most fruitful developments frequently take place at those points where two different lines of thought meet. These lines may have their roots in quite different parts of human culture, in different times or different cultural environments or different religious traditions: hence if they actually meet, that is, if they are at least so much related to each other that a real interaction can take place, then one may hope that new and interesting developments may follow.

Werner Heisenberg

THE TAO OF PHYSICS

PREFACE

Five years ago, I had a beautiful experience which set me on a road that has led to the writing of this book. I was sitting by the ocean one late summer afternoon, watching the waves rolling in and feeling the rhythm of my breathing, when I suddenly became aware of my whole environment as being engaged in a gigantic cosmic dance. Being a physicist, I knew that the sand, rocks, water, and air around me were made of vibrating molecules and atoms, and that these consisted of particles which interacted with one another by creating and destroying other particles. I knew also that the earth's atmosphere was continually bombarded by showers of "cosmic rays," particles of high energy undergoing multiple collisions as they penetrated the air. All this was familiar to me from my research in high-energy physics, but until that moment I had only experienced it through graphs, diagrams, and mathematical theories. As I sat on that beach my former experiences came to life; I "saw" cascades of energy coming down from outer space, in which particles were created and destroyed in rhythmic pulses; I "saw" the atoms of the elements and those of my body participating in this cosmic dance of energy; I felt its rhythm and I "heard" its sound, and at that moment I *knew* that this was the Dance of Shiva, the Lord of Dancers worshiped by the Hindus.

I had gone through a long training in theoretical physics and had done several years of research. At the same time, I had become very interested in Eastern mysticism and had begun to see the parallels to modern physics. I was particularly attracted to the puzzling aspects of Zen which reminded me of the puzzles in quantum theory. At first, however, relating the two was a purely intellectual exercise. To overcome the gap between rational, analytical thinking and the meditative experience of mystical truth, was, and still is, very difficult for me.

In the beginning, I was helped on my way by "power plants" which showed me how the mind can flow freely; how spiritual insights come on their own, without any effort, emerging from the depth of consciousness. I remember the first such experience. Coming, as it did, after years of detailed analytical thinking, it was so overwhelming that I burst into tears, at the same time, not unlike Castaneda, pouring out my impressions on to a piece of paper.

Later came the experience of the Dance of Shiva which I have tried to capture in the photomontage shown in Plate 7. It was followed by many similar experiences which helped me gradually to realize that a consistent view of the world is beginning to emerge from modern physics which is harmonious with ancient Eastern wisdom. I took many notes over the years, and wrote a few articles about the parallels I kept discovering, until I finally summarized my experiences in the present book.

This book is intended for the general reader with an interest in Eastern mysticism who need not necessarily know anything about physics. I have tried to present the main concepts and theories of modern physics without any mathematics and in nontechnical language, although a few paragraphs may still appear difficult to the layperson at first reading. The technical terms I had to introduce are all defined where they appear for the first time and are listed in the index at the end of the book.

I also hope to find among my readers many physicists with an interest in the philosophical aspects of physics, who have as yet not come in contact with the religious philosophies of the East. They will find that Eastern mysticism provides a consistent and beautiful philosophical framework which can accommodate our most advanced theories of the physical world.

As far as the contents of the book are concerned, the reader may feel a certain lack of balance between the presentation of scientific and mystical thought. Throughout the book, his or her understanding of physics should progress steadily, but a comparable progression in the understanding of Eastern mysticism may not occur. This seems unavoidable, as mysticism is, above all, an experience that cannot be learned from books. A deeper understanding of any mystical tradition can only be felt when one decides to become actively involved in it. All I can hope to do is to generate the feeling that such an involvement would be highly rewarding.

During the writing of this book, my own understanding of Eastern thought has deepened considerably. For this I am indebted to two men who come from the East. I am profoundly grateful to Phiroz Mehta for opening my eyes to many aspects of Indian mysticism, and to my T'ai Chi master Liu Hsiu Ch'i for introducing me to living Taoism.

It is impossible to mention the names of everyone—scientists, artists, students, and friends—who have helped me formulate my ideas in stimulating discussions. I feel, however, that I owe special thanks to Graham Alexander, Jonathan Ashmore, Stratford Caldecott, Lyn Gambles, Sonia Newby, Ray Rivers, Joël Scherk, George Sudarshan, and—last but not least—Ryan Thomas.

Finally, I am indebted to Mrs. Pauly Bauer-Ynnhof of Vienna for her generous financial support at a time when it was needed most.

London, *Fritjof Capra*
December 1974

I
THE WAY OF PHYSICS

Any path is only a path, and there is no affront, to oneself or to others, in dropping it if that is what your heart tells you. . . . Look at every path closely and deliberately. Try it as many times as you think necessary. Then ask yourself, and yourself alone, one question. . . . Does this path have a heart? If it does, the path is good; if it doesn't it is of no use.

Carlos Castaneda,
The Teachings of Don Juan

1
MODERN PHYSICS
A PATH WITH A HEART?

Modern physics has had a profound influence on almost all aspects of human society. It has become the basis of natural science, and the combination of natural and technical science has fundamentally changed the conditions of life on our earth, both in beneficial and detrimental ways. Today, there is hardly an industry that does not make use of the results of atomic physics, and the influence these have had on the political structure of the world through their application to atomic weaponry is well known. However, the influence of modern physics goes beyond technology. It extends to the realm of thought and culture where it has led to a deep revision in man's conception of the universe and his relation to it. The exploration of the atomic and subatomic world in the twentieth century has revealed an unsuspected limitation of classical ideas, and has necessitated a radical revision of many of our basic concepts. The concept of matter in subatomic physics, for example, is totally different from the traditional idea of a material substance in classical physics. The same is true for concepts like space, time, or cause and effect. These concepts, however, are fundamental to our outlook on the world around us and with their radical transformation our whole world-view has begun to change.

These changes, brought about by modern physics, have been widely discussed by physicists and by philosphers over the past decades, but very seldom has it been realized that they all seem to lead in the same direction, toward a view of the world which is very similar to the views held in Eastern mysticism. The concepts of modern physics often show surprising parallels to the ideas expressed in the religious philosophies of the Far East. Although these parallels have not, as yet, been discussed extensively, they have been noticed by some of the great physicists of our century when they came in contact with Far Eastern culture during their lecture tours to India, China and Japan. The following three quotations serve as examples:

> The general notions about human understanding . . . which are illustrated by discoveries in atomic physics are not in the nature of things wholly unfamiliar, wholly unheard of, or new. Even in our own culture they have a history, and in Buddhist and Hindu thought a more considerable and central place. What we shall find is an exemplification, an encouragement, and a refinement of old wisdom.[1]
>
> *Julius Robert Oppenheimer*

> For a parallel to the lesson of atomic theory . . . [we must turn] to those kinds of epistemological problems with which already thinkers like the Buddha and Lao Tzu have been confronted, when trying to harmonize our position as spectators and actors in the great drama of existence.[2]
>
> *Niels Bohr*

> The great scientific contribution in theoretical physics that has come from Japan since the last war may be an indication of a certain relationship between philosophical ideas in the tradition of the Far East and the philosophical substance of quantum theory.[3]
>
> *Werner Heisenberg*

The purpose of this book is to explore this relationship between the concepts of modern physics and the basic ideas in the philosophical and religious traditions of the Far East. We shall see how the two foundations of twentieth-century physics—quantum theory and relativity theory—both force us to see the world very much in the way a Hindu, Buddhist, or Taoist sees it, and how this similarity strengthens when we look at the recent attempts to combine these two theories in order to describe the phenomena of the submicroscopic world: the properties and interactions of the subatomic particles of which all matter is made. Here the parallels between modern physics and Eastern mysticism are most striking, and we shall often encounter statements where it is almost impossible to say whether they have been made by physicists or by Eastern mystics.

When I refer to "Eastern mysticism," I mean the religious philosophies of Hinduism, Buddhism, and Taoism. Although these comprise a vast number of subtly interwoven spiritual disciplines and philosophical systems, the basic features of their world-view are the same. This view is not limited to the East, but can be found to some degree in all mystically oriented philosophies. The argument of this book could therefore be phrased more generally by saying that modern physics leads us to a view of the world which is very similar to the views held by mystics of all ages and traditions. Mystical traditions are present in all religions, and mystical elements can be found in many schools of Western philosophy. The parallels to modern physics appear not only in the *Vedas* of Hinduism, in the *I Ching*, or in the Buddhist *sutras*, but also in the fragments of Heraclitus, in the Sufism of Ibn Arabi, or in the teachings of the Yaqui sorcerer Don Juan. The difference between Eastern and Western mysticism is that mystical schools have always played a marginal role in the West, whereas they constitute the mainstream of Eastern philosophical and religious thought. I shall therefore, for the sake of simplicity, talk about the "Eastern world view" and shall only occasionally mention other sources of mystical thought.

If physics leads us today to a world view which is essentially mystical, it returns, in a way, to its beginning, 2,500 years ago. It is interesting to follow the evolution of Western science along its spiral path, starting from the mystical philosophies of the early Greeks, rising and unfolding in an impressive development of intellectual thought that increasingly turned away from its mystical origins to develop a world view which is in sharp contrast to that of the Far East. In its most recent stages, Western science is finally overcoming this view and coming back to those of the early Greek and the Eastern philosophies. This time, however, it is not only based on intuition, but also on experiments of great precision and sophistication, and on a rigorous and consistent mathematical formalism.

The roots of physics, as of all Western science, are to be found in the first period of Greek philosophy in the sixth century B.C., in a culture where science, philosophy and religion were not separated. The sages of the Milesian school in Ionia were not concerned with such distinctions. Their aim was to discover the essential nature, or real constitution, of things which they called "physis." The term "physics" is derived from this Greek word and meant therefore, originally, the endeavor of seeing the essential nature of all things.

This, of course, is also the central aim of all mystics, and the philosophy of the Milesian school did indeed have a strong mystical flavor. The Milesians were called "hylozoists," or "those who think matter is alive," by the later Greeks, because they saw no distinction between animate and inanimate, spirit and matter. In fact, they did not even have a word for matter, since they saw all forms of existence as manifestations of the "physis," endowed with life and spirituality. Thus Thales declared all things to be full of gods and Anaximander saw the universe as a kind of organism which was supported by "pneuma," the cosmic breath, in the same way as the human body is supported by air.

The monistic and organic view of the Milesians was very close to that of ancient Indian and Chinese philosophy, and the parallels to Eastern thought are even stronger in the philosophy of Heraclitus of Ephesus. Heraclitus believed in a world of perpetual change, of eternal 'Becoming.' For him, all static Being was based on deception, and his universal principle was fire; a symbol for the continuous flow and change of all things. Heraclitus taught that all changes in the world arise from the dynamic and cyclic interplay of opposites, and he saw any pair of opposites as a unity. This unity, which contains and transcends all opposing forces, he called the Logos.

The split of this unity began with the Eleatic school, which assumed a Divine Principle standing above all gods and men. This principle was first identified with the unity of the universe, but was later seen as an intelligent and personal God who stands above the world and directs it. Thus began a trend of thought which led, ultimately, to the separation of spirit and matter and to a dualism which became characteristic of Western philosophy.

A drastic step in this direction was taken by Parmenides of Elea, who was in strong opposition to Heraclitus. He called his basic principle the Being and held that it was unique and invariable. He considered change to be impossible and regarded the changes we seem to perceive in the world as mere illusions of the senses. The concept of an indestructible substance as the subject of varying properties grew out of this philosophy and became one of the fundamental concepts of Western thought.

In the fifth century B.C., the Greek philosophers tried to overcome the sharp contrast between the views of Parmenides and Heraclitus. In order to reconcile the idea of unchangeable Being (of Parmenides) with that of eternal Becoming (of Heraclitus), they assumed that the Being is manifest in certain invariable substances, the mixture and separation of which gives rise to the changes in the world. This led to the concept of the atom, the smallest indivisible unit of matter, which found its clearest expression in the philosophy of Leucippus and Democritus. The Greek atomists drew a clear line between spirit and matter, picturing matter as being made of several "basic building blocks." These were purely passive and intrinsically dead particles moving in the void. The cause of their motion was not explained, but was often associated with external forces which were assumed to be of spiritual origin and fundamentally different from matter. In subsequent centuries, this image became an essential element of Western thought, of the dualism between mind and matter, between body and soul.

As the idea of a division between spirit and matter took hold, the philosophers turned their attention to the spiritual world, rather than the material, to the human soul and the problems of ethics. These questions were to occupy Western thought for more than two thousand years after the culmination of Greek science and culture in the fifth and fourth centuries B.C. The scientific knowledge of antiquity was systematized and organized by Aristotle, who created the scheme which was to be the basis of the Western view of the universe for two thousand years. But Aristotle himself believed that questions concerning the human soul and the contemplation of God's perfection were much more valuable than investigations of the material world. The reason the Aristotelian model of the universe remained unchallenged for so long was precisely this lack of interest in the material world, and the strong hold of the Christian church which supported Aristotle's doctrines throughout the Middle Ages.

Further development of Western science had to wait until the Renaissance, when men began to free themselves from the influence of Aristotle and the church and showed a new interest in nature. In the late fifteenth century, the study of nature was approached, for the first time, in a truly scientific spirit and experiments were undertaken to test speculative ideas. As this development was paralleled by a growing interest in mathematics, it finally led to the formulation of proper scientific theories, based on experiment and expressed in mathematical language. Galileo was the first to combine empirical knowledge with mathematics and is therefore seen as the father of modern science.

The birth of modern science was preceded and accompanied by a development of philosophical thought which led to an extreme formulation of the spirit/matter dualism. This formulation appeared in the seventeenth century in the philosophy of René Descartes who based his view of nature on a fundamental division into two separate and independent realms: that of mind (*res cogitans*), and that of matter (*res extensa*). The "Cartesian" division allowed scientists to treat matter as dead and completely separate from themselves, and to see the material world as a multitude of different objects assembled into a huge machine. Such a mechanistic world view was held by Isaac Newton, who constructed his mechanics on its basis and made it the foundation of classical physics. From the second half of the seventeenth to the end of the nineteenth century, the mechanistic Newtonian model of the universe dominated all scientific thought. It was paralleled by the image of a monarchical God who ruled the world from above by imposing his divine law on it. The fundamental laws of nature searched for by the scientists were thus seen as the laws of God, invariable and eternal, to which the world was subjected.

The philosophy of Descartes was not only important for the development of classical physics, but also had a tremendous influence on the general Western way of thinking up to the present day. Descartes's famous sentence "*Cogito ergo sum*"—"I think, therefore I exist"—has led Western man to equate his identity with his mind, instead of with his whole organism. As a consequence of the Cartesian division, most individuals are aware of themselves as isolated egos existing "inside" their bodies. The mind has been separated from the body and given the futile task of controlling it, thus causing an apparent conflict between the conscious will and the involuntary instincts. Each individual has been split up further into a large number of separate compartments, according to his or her activities, talents, feelings, beliefs, etc., which are engaged in endless conflicts generating continuous metaphysical confusion and frustration.

This inner fragmentation of man mirrors his view of the world "outside," which is seen as a multitude of separate objects and events. The natural environment is treated as if it consisted of separate parts to be exploited by different interest groups. The fragmented view is further extended to society, which is split into different nations, races, religious and political groups. The belief that all these fragments—in ourselves, in our environment, and in our society—are really separate can be seen as the essential reason for the present series of social, ecological, and cultural crises. It has alienated us from nature and from our fellow human beings. It has brought a grossly unjust distribution of natural resources, creating economic and political disorder; an ever-rising wave of violence, both spontaneous and institutionalized, and an ugly, polluted environment in which life has often become physically and mentally unhealthy.

The Cartesian division and the mechanistic world view have thus been beneficial and detrimental at the same time. They were extremely successful in the development of classical physics and technology, but had many adverse consequences for our civilization. It is fascinating to see that twentieth-century science, which originated in the Cartesian split and in the mechanistic world-view, and which indeed only became possible because of such a view, now overcomes this fragmentation and leads back to the idea of unity expressed in the early Greek and Eastern philosophies.

In contrast to the mechanistic Western view, the Eastern view of the world is "organic." For the Eastern mystic, all things and events perceived by the senses are interrelated, connected, and are but different aspects or manifestations of the same ultimate reality. Our tendency to divide the perceived world into individual and separate things and to experience ourselves as isolated egos in this world is seen as an illusion which comes from our measuring and categorizing mentality. It is called *avidya*, or ignorance, in Buddhist philosophy, and is seen as the state of a disturbed mind which has to be overcome:

> When the mind is disturbed, the multiplicity of things is produced, but when the mind is quieted, the multiplicity of things disappears.[4]

Although the various schools of Eastern mysticism differ in many details, they all emphasize the basic unity of the universe which is the central feature of their teachings. The highest aim for their followers—whether they are Hindus, Buddhists or Taoists—is to become aware of the unity and mutual interrelation of all things, to transcend the notion of an isolated individual self, and to identify themselves with the ultimate reality. The emergence of this awareness—known as 'enlightenment'—is not only an intellectual act, but is an experience which involves the whole person and is religious in its ultimate nature. For this reason, most Eastern philosophies are essentially religious philosophies.

In the Eastern view, then, the division of nature into separate objects is not fundamental and any such objects have a fluid and ever-changing character. The Eastern world view is therefore intrinsically dynamic and contains time and change as essential features. The cosmos is seen as one inseparable reality—forever in motion, alive, organic; spiritual and material at the same time.

Since motion and change are essential properties of things, the forces causing the motion are not outside the objects, as in the classical Greek view, but are an intrinsic property of matter. Correspondingly, the Eastern image of the Divine is not that of a ruler who directs the world from above, but of a principle that controls everything from within:

> He who, dwelling in all things,
> Yet is other than all things,
> Whom all things do not know,
> Whose body all things are,
> Who controls all things from within—
> He is your Soul, the Inner Controller,
> The Immortal.[5]

The following chapters will show that the basic elements of the Eastern world view are also those of the world-view emerging from modern physics. They are intended to suggest that Eastern thought—and, more generally, mystical thought—provides a consistent and relevant philosophical background to the theories of contemporary science; a conception of the world in which man's scientific discoveries can be in perfect harmony with his spiritual aims and religious beliefs. The two basic themes of this conception are the unity and interrelation of all phenomena and the intrinsically dynamic nature of the universe. The further we penetrate into the submicroscopic world, the more we shall realize how the modern physicist, like the Eastern mystic, has come to see the world as a system of inseparable, interacting, and ever-moving components, with man as an integral part of this system.

The organic, "ecological" world-view of the Eastern philosophies is no doubt one of the main reasons for the immense popularity they have recently gained in the West, especially among young people. In our Western culture, which is still dominated by the mechanistic, fragmented view of the world, an increasing number of people have seen this as the underlying reason for the widespread dissatisfaction in our society, and many have turned to Eastern ways of liberation. It is interesting, and perhaps not too surprising, that those who are attracted by Eastern mysticism, who consult the *I Ching* and practise Yoga or other forms of meditation, in general have a marked antiscientific attitude. They tend to see science, and physics in particular, as an unimaginative, narrow-minded discipline which is responsible for all the evils of modern technology.

This book aims at improving the image of science by showing that there is an essential harmony between the spirit of Eastern wisdom and Western science. It attempts to suggest that modern physics goes far beyond technology, that the way—or *Tao*—of physics can be a path with a heart, a way to spiritual knowledge and self-realization.

2
KNOWING AND SEEING

From the unreal lead me to the real!
From darkness lead me to light!
From death lead me to immortality!
Brihad-aranyaka Upanishad

Before studying the parallels between modern physics and Eastern mysticism, we have to deal with the question of how we can make any comparison at all between an exact science, expressed in the highly sophisticated language of modern mathematics, and spiritual disciplines which are mainly based on meditation and insist on the fact that their insights cannot be communicated verbally.

What we want to compare are the statements made by scientists and Eastern mystics about their knowledge of the world. To establish the proper framework for this comparison, we must first ask ourselves what kind of "knowledge" we are talking about; does the Buddhist monk from Angkor Wat or Kyoto mean the same thing by "knowledge" as the physicist from Oxford or Berkeley? Second, what kind of statements are we going to compare? What are we going to select from the experimental data, equations and theories on the one side, and from the religious scriptures, ancient myths, or philosophical treatises on the other? This chapter is intended to clarify these two points: the nature of the knowledge involved and the language in which this knowledge is expressed.

Throughout history, it has been recognized that the human mind is capable of two kinds of knowledge, or two modes of consciousness, which have often been termed the rational and the intuitive, and have traditionally been associated with science and religion, respectively. In the West, the intuitive, religious type of knowledge is often devalued in favor of rational, scientific knowledge, whereas the traditional Eastern attitude is in general just the opposite. The following statements about knowledge by two great minds of the West and the East typify the two positions. Socrates in Greece made the famous statement, "I know that I know nothing," and Lao Tzu in China said, "Not knowing that one knows is best." In the East, the values attributed to the two kinds of knowledge are often already apparent from the names given to them. The *Upanishads*, for example, speak about a higher and a lower knowledge and associate the lower knowledge with various sciences, the higher with religious awareness. Buddhists talk about 'relative' and 'absolute' knowledge, or about 'conditional truth' and 'transcendental truth.' Chinese philosophy, on the other hand, has always emphasized the complementary nature of the intuitive and the rational and has represented them by the archetypal pair *yin* and *yang* which form the basis of Chinese thought. Accordingly, two complementary philosophical traditions—Taoism and Confucianism—have developed in ancient China to deal with the two kinds of knowledge.

Rational knowledge is derived from the experience we have with objects and events in our everyday environment. It belongs to the realm of the intellect, whose function it is to discriminate, divide, compare, measure and categorize. In this way, a world of intellectual distinctions is created; of opposites which can exist only in relation to each other, which is why Buddhists call this type of knowledge 'relative.'

Abstraction is a crucial feature of this knowledge, because in order to compare and to classify the immense variety of shapes, structures, and phenomena around us we cannot take all their features into account, but have to select a few significant ones. Thus we construct an intellectual map of reality in which things are reduced to their general outlines. Rational knowledge is thus a system of abstract concepts and symbols, characterized by the linear, sequential structure which is typical of our thinking and speaking. In most languages this linear structure is made explicit by the use of alphabets which serve to communicate experience and thought in long lines of letters.

The natural world, on the other hand, is one of infinite varieties and complexities, a multidimensional world which contains no straight lines or completely regular shapes, where things do not happen in sequences, but all together; a world where—as modern physics tells us—even empty space is curved. It is clear that our abstract system of conceptual thinking can never describe or understand this reality completely. In thinking about the world we are faced with the same kind of problem as the cartographer who tries to cover the curved face of the earth with a sequence of plane maps. We can only expect an approximate representation of reality from such a procedure, and all rational knowledge is therefore necessarily limited.

The realm of rational knowledge is, of course, the realm of science which measures and quantifies, classifies and analyzes. The limitations of any knowledge obtained by these methods have become increasingly apparent in modern science, and in particular in modern physics which has taught us, in the words of Werner Heisenberg, "that every word or concept, clear as it may seem to be, has only a limited range of applicability."[1]

For most of us it is very difficult to be constantly aware of the limitations and of the relativity of conceptual knowledge. Because our representation of reality is so much easier to grasp than reality itself, we tend to confuse the two and to take our concepts and symbols for reality. It is one of the main aims of Eastern mysticism to rid us of this confusion. Zen Buddhists say that a finger is needed to point at the moon, but that we should not trouble ourselves with the finger once the moon is recognized; the Taoist sage Chuang Tzu wrote:

> Fishing baskets are employed to catch fish; but when the fish are got, the men forget the baskets; snares are employed to catch hares; but when the hares are got, men forget the snares. Words are employed

> to convey ideas; but when the ideas are grasped, men forget the words.[2]

In the West, the semanticist Alfred Korzybski made exactly the same point with his powerful slogan, "The map is not the territory."

What the Eastern mystics are concerned with is a direct experience of reality which transcends not only intellectual thinking but also sensory perception. In the words of the *Upanishads*,

> What is soundless, touchless, formless, imperishable,
> Likewise tasteless, constant, odorless,
> Without beginning, without end, higher than the great, stable—
> By discerning That, one is liberated from the mouth of death.[3]

Knowledge which comes from such an experience is called 'absolute knowledge' by Buddhists because it does not rely on the discriminations, abstractions, and classifications of the intellect which, as we have seen, are always relative and approximate. It is, so we are told by Buddhists, the direct experience of undifferentiated, undivided, indeterminate 'suchness.' Complete apprehension of this suchness is not only the core of Eastern mysticism, but is the central characteristic of all mystical experience.

The Eastern mystics repeatedly insist on the fact that the ultimate reality can never be an object of reasoning or of demonstrable knowledge. It can never be adequately described by words because it lies beyond the realms of the senses and of the intellect from which our words and concepts are derived. The *Upanishads* say about it:

> There the eye goes not,
> Speech goes not, nor the mind.
> We know not, we understand not
> How one would teach it.[4]

Lao Tzu, who calls this reality the *Tao*, states the same fact in the opening line of the *Tao Te Ching*: "The *Tao* that can be expressed is not the eternal *Tao*." The fact—obvious from any reading of the newspapers—that mankind has not become much wiser over the past two thousand years, in spite of a prodigious increase in rational knowledge, is ample evidence of the impossibility of communicating absolute knowledge by words. As Chuang Tzu said, "If it could be talked about, everybody would have told their brother."[5]

Absolute knowledge is thus an entirely nonintellectual experience of reality, an experience arising in a non-ordinary state of consciousness which may be called a 'meditative' or mystical state. That such a state exists has not only been testified by numerous mystics in the East and West but is also indicated by psychological research. In the words of William James:

> Our normal waking consciousness, rational consciousness as we call it, is but one special type of consciousness, whilst all about it, parted from it by the filmiest of screens, there lie potential forms of consciousness entirely different.[6]

Although physicists are mainly concerned with rational knowledge and mystics with intuitive knowledge, both types of knowledge occur in both fields. This becomes apparent when we examine how knowledge is obtained and how it is expressed, both in physics and Eastern mysticism.

In physics, knowledge is acquired through the process of scientific research which can be seen to proceed in three stages. The first stage consists in gathering experimental evidence about the phenomena to be explained. In the second stage, the experimental facts are correlated with mathematical symbols and a mathematical scheme is worked out which interconnects these symbols in a precise and consistent way. Such a scheme is usually called a mathematical model or, if it is more comprehensive, a theory. This theory is then used to predict the results of further experiments which are undertaken to check all its implications. At this stage, physicists may be satisfied when they have found a mathematical scheme and know how to use it to predict experiments. But eventually, they will want to talk about their results to nonphysicists and will therefore have to express them in plain language. This means they will have to formulate a model in ordinary language which interprets their mathematical scheme. Even for the physicists themselves, the formulation of such a verbal model, which constitutes the third stage of research, will be a criterion of the understanding they have reached.

In practice, of course, the three stages are not neatly separated and do not always occur in the same order. For example, a physicist may be led to a particular model by some philosophical belief he (or she) holds, which he may continue to believe in, even when contrary experimental evidence arises. He will then—and this happens in fact very often—try to modify his model so that it can account for the new experiments. But if experimental evidence continues to contradict the model, he will eventually be forced to drop it.

This way of basing all theories firmly on experiment is known as the scientific method, and we shall see that it has its counterpart in Eastern philosophy. Greek philosophy, on the other hand, was fundamentally different in that respect. Although Greek philosophers had extremely ingenious ideas about nature which often come very close to modern scientific models, the enormous difference between the two is the empirical attitude of modern science which was by and large foreign to the Greek mind. The Greeks obtained their models deductively from some fundamental axiom or principle and not inductively from what had been observed. On the other hand, of course, the Greek art of deductive reasoning and logic is an essential ingredient in the second stage of scientific research, the formulation of a consistent mathematical model, and thus an essential part of science.

Rational knowledge and rational activities certainly constitute the major part of scientific research, but are not all there is to it. The rational part of research would, in fact, be useless if it were not complemented by the intuition that gives scientists new insights and makes them creative. These insights tend to come suddenly and, characteristically, not when sitting at a desk working out the equations, but when relaxing in the bath, during a walk in the woods, on the beach, etc. During these periods of relaxation after concentrated intellectual activity, the intuitive mind seems to take over and can produce the sudden clarifying insights which give so much joy and delight to scientific research.

Intuitive insights, however, are of no use to physics unless they can be formulated in a consistent mathematical framework, supplemented by an interpretation in plain language. Abstraction is a crucial feature of this framework. It consists, as mentioned before, of a system of concepts and symbols which constitute a map of reality. This map represents only some features of reality; we do not know exactly which these are, since we started compiling our map gradually and without critical analysis in our childhood. The words of our language are thus not clearly defined. They have several meanings, many of which pass only vaguely through our mind and remain largely in our subconscious when we hear a word.

The inaccuracy and ambiguity of our language is essential for poets who work largely with its subconscious layers and associations. Science, on the other hand, aims for clear definitions and unambiguous connections, and therefore it abstracts language further by limiting the meaning of its words and by standardizing its structure, in accordance with the rules of logic. The ultimate abstraction takes place in mathematics where words are replaced by symbols and where the operations of connecting the symbols are rigorously defined. In this way, scientists can condense information into one equation, i.e. into one single line of symbols, for which they would need several pages of ordinary writing.

The view that mathematics is nothing but an extremely abstracted and compressed language does not go unchallenged. Many mathematicians, in fact, believe that mathematics is not just a language to describe nature, but is inherent in nature itself. The originator of this belief was Pythagoras who made the famous statement, "All things are numbers," and developed a very special kind of mathematical mysticism. Pythagorean philosophy thus introduced logical reasoning into the domain of religion, a development which, according to Bertrand Russell, was decisive for Western religious philosophy:

> The combination of mathematics and theology, which began with Pythagoras, characterized religious philosophy in Greece, in the Middle Ages, and in modern times down to Kant. . . . In Plato, St. Augustine, Thomas Aquinas, Descartes, Spinoza,

> and Leibniz there is an intimate blending of religion and reasoning, of moral aspiration with logical admiration of what is timeless, which comes from Pythagoras, and distinguishes the intellectualized theology of Europe from the more straightforward mysticism of Asia.[7]

The "more straightforward mysticism of Asia" would, of course, not adopt the Pythagorean view of mathematics. In the Eastern view, mathematics, with its highly differentiated and well defined structure, must be seen as part of our conceptual map and not as a feature of reality itself. Reality, as experienced by the mystic, is completely indeterminate and undifferentiated.

The scientific method of abstraction is very efficient and powerful, but we have to pay a price for it. As we define our system of concepts more precisely, as we streamline it and make the connections more and more rigorous, it becomes increasingly detached from the real world. Using again Korzybski's analogy of the map and the territory, we could say that ordinary language is a map which, due to its intrinsic inaccuracy, has a certain flexibility so that it can follow the curved shape of the territory to some degree. As we make it more rigorous, this flexibility gradually disappears, and with the language of mathematics we have reached a point where the links with reality are so tenuous that the relation of the symbols to our sensory experience is no longer evident. This is why we have to supplement our mathematical models and theories with verbal interpretations, again using concepts which can be understood intuitively, but which are slightly ambiguous and inaccurate.

It is important to realize the difference between the mathematical models and their verbal counterparts. The former are rigorous and consistent as far as their internal structure is concerned, but their symbols are not directly related to our experience. The verbal models, on the other hand, use concepts which can be understood intuitively, but are always inaccurate and ambiguous. They are in this respect not different from philosophical models of reality, and thus the two can very well be compared.

If there is an intuitive element in science, there is also a rational element in Eastern mysticism. The degree to which reason and logic are emphasized, however, varies enormously from one school to the other. The Hindu Vedanta, or the Buddhist Madhyamika, for example, are highly intellectual schools, whereas Taoists have always had a deep mistrust of reason and logic. Zen, which grew out of Buddhism but was strongly influenced by Taoism, prides itself on being "without words, without explanations, without instructions, without knowledge." It concentrates almost entirely on the experience of enlightenment and is only marginally interested in interpreting this experience. A well-known Zen phrase says, "The instant you speak about a thing, you miss the mark."

Although other schools of Eastern mysticism are less extreme, the direct mystical experience is at the core of all of them. Even those mystics who are engaged in the most sophisticated argumentation never see the intellect as their source of knowledge but use it merely to analyze and interpret their personal mystical experience. All knowledge is firmly based on this experience, thus giving the Eastern traditions a strong empirical character that is always emphasized by its proponents. D. T. Suzuki, for example, writes of Buddhism:

> Personal experience is . . . the foundation of Buddhist philosophy. In this sense Buddhism is radical empiricism or experientialism, whatever dialectic later developed to probe the meaning of enlightenment-experience.[8]

Joseph Needham repeatedly brings the empirical attitude of Taoists into prominence in his work *Science and Civilization in China* and finds that this attitude has made Taoism the basis of Chinese science and technology. The early Taoist philosophers, in Needham's words, "withdrew into the wilderness, the forests and mountains, there to meditate upon the Order of Nature, and to observe its innumerable manifestations."[9] The same spirit is reflected in the Zen verses,

> He who would understand the meaning of Buddha-
> nature

> Must watch for the season and the causal relations.[10]

The firm basis of knowledge on experience in Eastern mysticism suggests a parallel to the firm basis of scientific knowledge on experiment. This parallel is further enforced by the nature of the mystical experience. It is described in the Eastern traditions as a direct insight which lies outside the realm of the intellect and is obtained by watching rather than thinking; by looking inside oneself; by observation.

In Taoism, this notion of observation is embodied in the name for Taoist temples, *kuan*, which originally meant "to look." Taoists thus regarded their temples as places of observation. In Ch'an Buddhism, the Chinese version of Zen, enlightenment is often referred to as 'the vision of the Tao,' and seeing is regarded as the basis of knowing in all Buddhist schools. The first item of the Eightfold Path, the Buddha's prescription for self-realization, is right seeing, followed by right knowing. D. T. Suzuki writes on this point:

> The seeing plays the most important role in Buddhist epistemology, for seeing is at the basis of knowing. Knowing is impossible without seeing; all knowledge has its origin in seeing. Knowing and seeing are thus found generally united in Buddha's teaching. Buddhist philosophy therefore ultimately points to seeing reality as it is. Seeing is experiencing enlightenment.[11]

This passage is also reminiscent of the Yaqui mystic Don Juan, who says, "My predilection is to see . . . because only by seeing can a man of knowledge know."[12]

A word of caution should perhaps be added here. The emphasis on seeing in mystical traditions should not be taken too literally, but has to be understood in a metaphorical sense, since the mystical experience of reality is an essentially nonsensory experience. When the Eastern mystics talk about 'seeing,' they refer to a mode of perception which may include visual perception, but which always and essentially transcends it to become a nonsensory experience of reality. What they do emphasize, however, when they talk about seeing, looking or observing, is the empirical character of their knowledge. This empirical approach of Eastern philosophy is strongly reminiscent of the emphasis on observation in science and thus suggests a framework for our comparison. The experimental stage in scientific research seems to correspond to the direct insight of the Eastern mystic, and the scientific models and theories correspond to the various ways in which this insight is interpreted.

The parallel between scientific experiments and mystical experiences may seem surprising in view of the very different nature of these acts of observation. Physicists perform experiments involving an elaborate teamwork and a highly sophisticated technology, whereas mystics obtain their knowledge purely through introspection, without any machinery, in the privacy of meditation. Scientific experiments, furthermore, seem to be repeatable any time and by anybody, whereas mystical experiences seem to be reserved for a few individuals at special occasions. A closer examination shows, however, that the differences between the two kinds of observation lie only in their approach and not in their reliability or complexity.

Anybody who wants to repeat an experiment in modern subatomic physics has to undergo many years of training. Only then will he or she be able to ask nature a specific question through the experiment and to understand the answer. Similarly, a deep mystical experience requires, generally, many years of training under an experienced master and, as in the scientific training, the dedicated time does not alone guarantee success. If the student is successful, however, he or she will be able to "repeat the experiment." The repeatability of the experience is, in fact, essential to every mystical training and is the very aim of the mystics' spiritual instruction.

A mystical experience, therefore, is not any more unique than a modern experiment in physics. On the other hand, it is not less sophisticated either, although its sophistication is of a very different kind. The complexity and efficiency of the physicist's technical apparatus is matched, if not surpassed, by that of the mystic's consciousness—both physical and spiritual—in deep meditation. The scientists and the mystics, then, have developed highly sophisticated methods of observing nature which are inaccessible to the layperson. A page from a journal of modern experimental physics will be as mysterious to the uninitiated as a Tibetan mandala. Both are records of inquiries into the nature of the universe.

Although deep mystical experiences do not, in general, occur without long preparation, direct intuitive insights are experienced by all of us in our everyday lives. We are all familiar with the situation where we have forgotten the name of a person or place, or some other word, and cannot produce it in spite of the utmost concentration. We have it "on the tip of our tongue," but it just will not come out, until we give up and shift our attention to something else when suddenly, in a flash, we remember the forgotten name. No thinking is involved in this process. It is a sudden, immediate insight. This example of suddenly remembering something is particularly relevant to Buddhism, which holds that our original nature is that of the enlightened Buddha and that we have just forgotten it. Students of Zen Buddhism are asked to discover their 'original face' and the sudden 'remembering' of this face is their enlightenment.

Another well-known example of spontaneous intuitive insights is the joke. In the split-second where you understand a joke, you experience a moment of 'enlightenment.' It is well known that this moment must come spontaneously, that it cannot be achieved by "explaining" the joke, i.e. by intellectual analysis. Only with a sudden intuitive insight into the nature of the joke do we experience the liberating laughter the joke is meant to produce. The similarity between a spiritual insight and the understanding of a joke must be well known to enlightened men and women, since they almost invariably show a great sense of humor. Zen, especially, is full of funny stories and anecdotes, and in the *Tao Te Ching* we read, "If it were not laughed at, it would not be sufficient to be *Tao*."[13]

In our everyday life, direct intuitive insights into the nature of things are normally limited to extremely brief moments. Not so in Eastern mysticism where they are extended to long periods and, ultimately, become a constant awareness. The preparation of the mind for this awareness—for the immediate, nonconceptual awareness of reality—is the main purpose of all schools of Eastern mysticism, and of many aspects of the Eastern way of life. During the long cultural history of India, China, and Japan, an enormous variety of techniques, rituals and art forms have been developed to achieve this purpose, all of which may be called meditation in the widest sense of the word.

The basic aim of these techniques seems to be to silence the thinking mind and to shift the awareness from the rational to the intuitive mode of consciousness. In many forms of meditation, this silencing of the rational mind is achieved by concentrating one's attention on a single item, like one's breathing, the sound of a mantra, or the visual image of a mandala. Other schools focus the attention on body movements which have to be performed spontaneously without the interference of any thought. This is the way of the Hindu Yoga and of the Taoist *T'ai Chi Ch'uan*. The rhythmical movements of these schools can lead to the same feeling of peace and serenity which is characteristic of the more static forms of meditation; a feeling which, incidentally, may be evoked also by some sports. In my experience, for example, skiing has been a highly rewarding form of meditation.

Eastern art forms, too, are forms of meditation. They are not so much means for expressing the artist's ideas as ways of self-realization through the development of the intuitive mode of consciousness. Indian music is not learned by reading notes, but by listening to the playing of the teacher and thus developing a feeling for the music, just as the *T'ai Chi* movements are not learned by verbal instructions but by doing them over and over again in unison with the teacher. Japanese tea ceremonies are full of slow, ritualistic movements. Chinese calligraphy requires the uninhibited, spontaneous movement of the hand. All these skills are used in the East to develop the meditative mode of consciousness.

For most people, and especially for intellectuals, this mode of consciousness is a completely new experience. Scientists are familiar with direct intuitive insights from their research, because every new discovery originates in such a sudden nonverbal flash. But these are extremely short moments which arise when the mind is filled with information, with concepts and thought patterns. In meditation, on the other hand, the mind is emptied of all thoughts and concepts and thus prepared to function for long periods through its intuitive mode. Lao Tzu speaks about this contrast between research and meditation when he says:

> He who pursues learning will increase every day;
> He who pursues *Tao* will decrease every day.[14]

When the rational mind is silenced, the intuitive mode produces an extraordinary awareness; the environment is experienced in a direct way without the filter of conceptual thinking. In the words of Chuang Tzu, "The still mind of the sage is a mirror of heaven and earth—the glass of all things."[15] The experience of oneness with the surrounding environment is the main characteristic of this meditative state. It is a state of consciousness where every form of fragmentation has ceased, fading away into undifferentiated unity.

In deep meditation, the mind is completely alert. In addition to the nonsensory apprehension of reality it also takes in all the sounds, sights, and other impressions of the surrounding environment, but it does not hold the sensory images to be analyzed or interpreted. They are not allowed to distract the attention. Such a state of awareness is not unlike the state of mind of a warrior who expects an attack in extreme alertness, registering everything that goes on around him without being distracted by it for an instant. The Zen master Yasutani Roshi uses this image in his description of *shikan-taza*, the practice of Zen meditation:

> *Shikan-taza* is a heightened state of concentrated awareness wherein one is neither tense nor hurried, and certainly never slack. It is the mind of somebody facing death. Let us imagine that you are engaged in a duel of swordsmanship of the kind that used to take place in ancient Japan. As you face your opponent you are unceasingly watchful, set, ready. Were you to relax your vigilance even momentarily, you would be cut down instantly. A crowd gathers to see the fight. Since you are not blind you see them from the corner of your eye, and since you are not deaf you hear them. But not

> for an instant is your mind captured by these sense impressions.[16]

Because of the similarity between the meditative state and the frame of mind of a warrior, the image of the warrior plays an important role in the spiritual and cultural life of the East. The stage for India's favorite religious text, the *Bhagavad Gita*, is a battlefield, and martial arts constitute an important part in the traditional cultures of China and Japan. In Japan, the strong influence of Zen on the tradition of the samurai gave rise to what is known as *bushido*, "the way of the warrior," an art of swordsmanship where the spiritual insight of the swordsman reaches its highest perfection. The Taoist *T'ai Chi Ch'uan*, which was considered to be the supreme martial art in China, combines slow and rhythmical "yogic" movements with the total alertness of the warrior's mind in a unique way.

Eastern mysticism is based on direct insights into the nature of reality, and physics is based on the observation of natural phenomena in scientific experiments. In both fields, the observations are then interpreted, and the interpretation is very often communicated by words. Since words are always an abstract, approximate map of reality, the verbal interpretations of a scientific experiment or of a mystical insight are necessarily inaccurate and incomplete. Modern physicists and Eastern mystics alike are well aware of this fact.

In physics, the interpretations of experiments are called models or theories, and the realization that all models and theories are approximate is basic to modern scientific research. Thus the aphorism of Einstein, "As far as the laws of mathematics refer to reality, they are not certain; and as far as they are certain, they do not refer to reality." Physicists know that their methods of analysis and logical reasoning can never explain the whole realm of natural phenomena at once, and so they single out a certain group of phenomena and try to build a model to describe this group. In doing so, they neglect other phenomena, and the model will therefore not give a complete description of the real situation. The phenomena which are not taken into account may either have such a small effect that their inclusion would not alter the theory significantly, or they may be left out simply because they are not known at the time when the theory is built.

To illustrate these points, let us look at one of the best known models in physics, Newton's "classical" mechanics. The effects of air resistance or friction, for example, are generally not taken into account in this model because they are usually very small. But apart from such omissions, Newtonian mechanics was for a long time considered to be the final theory for the description of all natural phenomena, until electric and magnetic phenomena, which had no place in Newton's theory, were discovered. The discovery of these phenomena showed that the model was incomplete, that it could be applied only to a limited group of phenomena—essentially the motion of solid bodies.

Studying a limited group of phenomena can also mean studying their physical properties only over a limited range, which may be another reason for the theory to be approximate. This aspect of the approximation is quite subtle because we never know beforehand where the limitations of a theory lie. Only experience can tell. Thus the image of classical mechanics was further eroded when twentieth-century physics showed its essential limitations. Today we know that the Newtonian model is valid only for objects consisting of large numbers of atoms, and only for velocities which are small compared to the speed of light. When the first condition is not given, classical mechanics has to be replaced by quantum theory; when the second condition is not satisfied, relativity theory has to be applied. This does not mean that Newton's model is "wrong," or that quantum theory and relativity theory are "right." All these models are approximations which are valid for a certain range of phenomena. Beyond this range, they no longer give a satisfactory description of nature, and new models have to be found to replace the old ones—or, better, to extend them by improving the approximation.

To specify the limitations of a given model is often one of the most difficult, and yet one of the most important tasks in its construction. According to Geoffrey Chew, whose "bootstrap models" will be discussed at great length later on, it is essential that one should always ask, as soon as a certain model or theory is found to work: Why does it work? What are the model's limits? In what way, exactly, is it an approximation? These questions are seen by Chew as the first step toward further progress.

The Eastern mystics, too, are well aware of the fact that all verbal descriptions of reality are inaccurate and incomplete. The direct experience of reality transcends the realm of thought and language, and, since all mysticism is based on such a direct experience, everything that is said about it can only be partly true. In physics, the approximate nature of all statements is quantified and progress is made by improving the approximations in many successive steps. How, then, do the Eastern traditions deal with the problem of verbal communication?

First of all, mystics are mainly interested in the experience of reality and not in the description of this experience. They are therefore generally not interested in the analysis of such a description, and the concept of a well-defined approximation has thus never arisen in Eastern thought. If, on the other hand, Eastern mystics want to communicate their experience, they are confronted with the limitations of language. Several different ways have been developed in the East to deal with this problem.

Indian mysticism—and Hinduism, in particular—clothes its statements in the form of myths, using metaphors and symbols, poetic images, similes, and allegories. Mythical language is much less restricted by logic and common sense. It is full of magic and of paradoxical situations, rich in suggestive images and never precise, and can thus convey the way in which mystics experience reality much better than factual language. According to Ananda Coomaraswamy, "Myth embodies the nearest approach to absolute truth that can be stated in words."[17]

The rich Indian imagination has created a vast number of gods and goddesses whose incarnations and exploits are the subjects of fantastic tales, collected in epics of huge dimensions. The Hindu with deep insight knows that all these gods are creations of the mind, mythical images representing the many faces of reality. On the other hand, he also knows that they were not merely created to make the stories more attractive, but are essential vehicles to convey the doctrines of a philosophy rooted in mystical experience.

Chinese and Japanese mystics have found a different way of dealing with the language problem. Instead of making the paradoxical nature of reality palatable through the symbols and images of myth, they prefer very often to accentuate it by using factual language. Thus Taoists made frequent use of paradoxes in order to expose the inconsistencies arising from verbal communication and to show its limits. They have passed this technique on to Chinese and Japanese Buddhists who have developed it further. It has reached its extreme in Zen Buddhism with the so-called *koans*, those nonsensical riddles which are used by many Zen masters to transmit the teachings. These *koans* establish an important parallel to modern physics which will be taken up in the next chapter.

In Japan, there exists yet another mode of expressing philosophical views which should be mentioned. It is a special form of extremely concise poetry which is often used by Zen masters to point directly at the 'suchness' of reality. When a monk asked Fuketsu Ensho, "When speech and silence are both inadmissible, how can one pass without error?" the master replied:

> I always remember Kiangsu in March—
> The cry of the partridge,
> The mass of fragrant flowers.[18]

This form of spiritual poetry has reached its perfection in the *haiku*, a classical Japanese verse of just seventeen syllables, which was deeply influenced by Zen. The insight into the very nature of life reached by these haiku poets comes across even in the English translation:

> Leaves falling
> Lie on one another;
> The rain beats the rain.[19]

Whenever the Eastern mystics express their knowledge in words—be it with the help of myths, symbols, poetic images or paradoxical statements—they are well aware of the limitations imposed by language and "linear" thinking. Modern physics has come to take exactly the same attitude with regard to its verbal models and theories. They, too, are only approximate and necessarily inaccurate. They are the counterparts of the Eastern myths, symbols, and poetic images, and it is at this level that I shall draw the parallels. The same idea about matter is conveyed, for example, to the Hindu by the cosmic dance of the god Shiva as to the physicist by certain aspects of quantum-field theory. Both the dancing god and the physical theory are creations of the mind: models to describe their authors' intuition of reality.

3
BEYOND LANGUAGE

> The contradiction so puzzling to the ordinary way of thinking comes from the fact that we have to use language to communicate our inner experience which in its very nature transcends linguistics.
>
> *D. T. Suzuki*

> The problems of language here are really serious. We wish to speak in some way about the structure of the atoms. . . . But we cannot speak about atoms in ordinary language.
>
> *W. Heisenberg*

The notion that all scientific models and theories are approximate and that their verbal interpretations always suffer from the inaccuracy of our language was already commonly accepted by scientists at the beginning of this century, when a new and completely unexpected development took place. The study of the world of atoms forced physicists to realize that our common language is not only inaccurate, but totally inadequate to describe the atomic and subatomic reality. Quantum theory and relativity theory, the two bases of modern physics, have made it clear that this reality transcends classical logic and that we cannot talk about it in ordinary language. Thus Heisenberg writes:

> The most difficult problem . . . concerning the use of the language arises in quantum theory. Here we

> have at first no simple guide for correlating the mathematical symbols with concepts of ordinary language; and the only thing we know from the start is the fact that our common concepts cannot be applied to the structure of the atoms.[1]

From a philosophical point of view, this has certainly been the most interesting development in modern physics, and here lies one of the roots of its relation to Eastern philosophy. In the schools of Western philosophy, logic and reasoning have always been the main tools used to formulate philosophical ideas and this is true, according to Bertrand Russell, even of religious philosophies. In Eastern mysticism, on the other hand, it has always been realized that reality transcends ordinary language, and the sages of the East were not afraid to go beyond logic and common concepts. This is the main reason, I think, why their models of reality constitute a more appropriate philosophical background to modern physics than the models of Western philosophy.

The problem of language encountered by the Eastern mystic is exactly the same as the problem the modern physicist faces. In the two passages quoted at the beginning of this chapter, D. T. Suzuki speaks about Buddhism[2] and Werner Heisenberg speaks about atomic physics,[3] and yet the two passages are almost identical. Both the physicist and the mystic want to communicate their knowledge, and when they do so with words their statements are paradoxical and full of logical contradictions. These paradoxes are characteristic of all mysticism, from Heraclitus to Don Juan, and since the beginning of this century they are also characteristic of physics.

In atomic physics, many of the paradoxical situations are connected with the dual nature of light or—more generally—of electromagnetic radiation. On the one hand, it is clear that this radiation must consist of waves because it produces the well-known interference phenomena associated with waves: when there are two sources of light, the intensity of the light to be found at some other place will not necessarily be just the sum of that which comes from the two sources, but may be more or less. This can easily be explained by the interference of the waves emanating from the two sources: in those places where two crests coincide, we shall have more light than the sum of the two; where a crest and a trough coincide, we shall have less. The precise amount of interference can easily be calculated. Interference phenomena of this

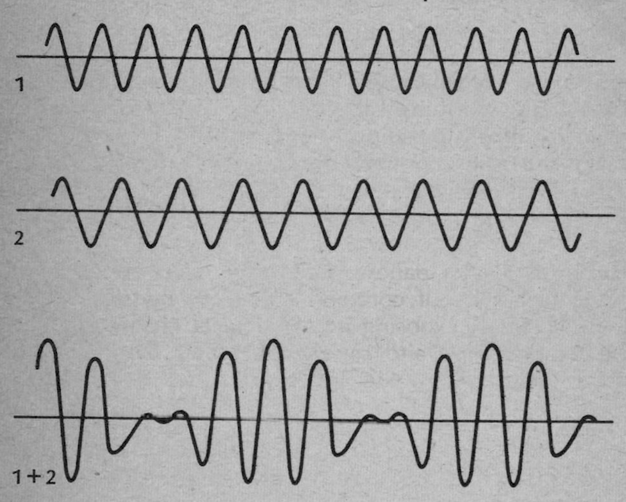

interference of two waves

kind can be observed whenever one deals with electromagnetic radiation, and force us to conclude that this radiation consists of waves.

On the other hand, electromagnetic radiation also produces the so-called photoelectric effect: when ultraviolet light is shone on the surface of some metals it can "kick out" electrons from the surface of the metal, and therefore it must consist of moving particles. A similar situation occurs in the "scattering" experiments of X rays. These experiments can be interpreted correctly only if they are described as collisions of "light particles" with electrons. And yet, they show the interference patterns characteristic of waves. The question which puzzled physicists so much in the early stages of atomic theory was how electromagnetic radiation could simultaneously consist of particles (i.e., of entities confined to a very small volume) and of waves, which are spread out over a large area of space. Neither language nor imagination could deal with this kind of reality very well.

Eastern mysticism has developed several different ways of dealing with the paradoxical aspects of reality. Whereas they are bypassed in Hinduism through the use of mythical language, Buddhism and Taoism tend to emphasize the paradoxes rather than conceal them. The main Taoist scripture, Lao Tzu's *Tao Te Ching*, is written in an extremely puzzling, seemingly illogical style. It is full of intriguing contradictions and its compact, powerful, and extremely poetic language is meant to arrest the reader's mind and throw it off its familiar tracks of logical reasoning.

Chinese and Japanese Buddhists have adopted this Taoist technique of communicating the mystical experience by simply exposing its paradoxical character. When the Zen master Daito saw the Emperor Godaigo, who was a student of Zen, the master said:

> We were parted many thousands of *kalpas* ago, yet we have not been separated even for a moment. We are facing each other all day long, yet we have never met.[4]

Zen Buddhists have a particular knack for making a virtue out of the inconsistencies arising from verbal communication, and with the *koan* system they have developed a unique way of transmitting their teachings completely nonverbally. *Koans* are carefully devised nonsensical riddles which are meant to make the student of Zen realize the limitations of logic and reasoning in the most dramatic way. The irrational wording and paradoxical content of these riddles makes it impossible to solve them by thinking. They are designed precisely to stop the thought process and thus to make the student ready for the nonverbal experience of reality. The contemporary Zen master Yasutani introduced a Western student to one of the most famous *koans* with the following words:

> One of the best *koans*, because the simplest, is *Mu*. This is its background: A monk came to Joshu, a renowned Zen master in China hundreds of years

> ago, and asked: "Has a dog Buddha-nature or not?" Joshu retorted, "*Mu!*" Literally, the expression means "no" or "not," but the significance of Joshu's answer does not lies in this. *Mu* is the expression of the living, functioning, dynamic Buddha-nature. What you must do is discover the spirit or essence of this *Mu*, not through intellectual analysis but by search into your innermost being. Then you must demonstrate before me, concretely and vividly, that you understand *Mu* as living truth, without recourse to conceptions, theories, or abstract explanations. Remember, you can't understand *Mu* through ordinary cognition; you must grasp it directly with your whole being.[5]

To a beginner, the Zen master will normally present either this *Mu-koan* or one of the following two:

> "What was your original face—the one you had before your parents gave birth to you?"

> "You can make the sound of two hands clapping. Now what is the sound of one hand?"

All these *koans* have more or less unique solutions which a competent master recognizes immediately. Once the solution is found, the *koan* ceases to be paradoxical and becomes a profoundly meaningful statement made from the state of consciousness which it has helped to awaken.

In the Rinzai school, the student has to solve a long series of *koans*, each of them dealing with a particular aspect of Zen. This is the only way this school transmits its teachings. It does not use any positive statements, but leaves it entirely to the student to grasp the truth through the *koans*.

Here we find a striking parallel to the paradoxical situations which confronted physicists at the beginning of atomic physics. As in Zen, the truth was hidden in paradoxes that could not be solved by logical reasoning, but had to be understood in the terms of a new awareness; the awareness of the atomic reality. The teacher here was, of course, nature, who, like the Zen masters, does not provide any statements. She just provides the riddles.

The solving of a *koan* demands a supreme effort of concentration and involvement from the student. In books about Zen we read that the *koan* grips the student's heart and mind and creates a true mental impasse, a state of sustained tension in which the whole world becomes an enormous mass of doubt and questioning. The founders of quantum theory experienced exactly the same situation, described here most vividly by Heisenberg:

> I remember discussions with Bohr which went through many hours till very late at night and ended almost in despair; and when at the end of the discussion I went alone for a walk in the neighboring park I repeated to myself again and again the question: Can nature possibly be so absurd as it seemed to us in these atomic experiments?[6]

Whenever the essential nature of things is analyzed by the intellect, it must seem absurd or paradoxical. This has always been recognized by the mystics, but has become a problem in science only very recently. For centuries, scientists were searching for the "fundamental laws of nature" underlying the great variety of natural phenomena. Those phenomena belonged to the scientists' macroscopic environment and thus to the realm of their sensory experience. Since the images and intellectual concepts of their language were abstracted from this very experience, they were sufficient and adequate to describe the natural phenomena.

Questions about the essential nature of things were answered in classical physics by the Newtonian mechanistic model of the universe which, much in the same way as the Democritean model in ancient Greece, reduced all phenomena to the motions and interactions of hard, indestructible atoms. The properties of these atoms were abstracted from the macroscopic notion of billiard balls, and thus from sensory experience. Whether this notion could actually be applied to the world of atoms was not questioned. Indeed, it could not be investigated experimentally.

In the twentieth century, however, physicists were able to tackle the question about the ultimate nature of matter experimentally. With the help of a most sophisticated technology they were able to probe deeper and deeper into nature, uncovering one layer of matter after the other in search for its ultimate "building blocks." Thus the existence of atoms was verified, then their constituents were discovered—the nuclei and electrons—and finally the components of the nucleus—the protons and neutrons—and many other subatomic particles.

The delicate and complicated instruments of modern experimental physics penetrate deep into the submicroscopic world, into realms of nature far removed from our macroscopic environment, and make this world accessible to our senses. However, they can do so only through a chain of processes ending, for example, in the audible click of a Geiger counter, or in a dark spot on a photographic plate. What we see, or hear, are never the investigated phenomena themselves but always their consequences. The atomic and subatomic world itself lies beyond our sensory perception.

It is, then, with the help of modern instrumentation that we are able to "observe" the properties of atoms and their constituents in an indirect way, and thus to "experience" the subatomic world to some extent. This experience, however, is not an ordinary one, comparable to that of our daily environment. The knowledge about matter at this level is no longer derived from direct sensory experience, and therefore our ordinary language, which takes its images from the world of the senses, is no longer adequate to describe the observed phenomena. As we penetrate deeper and deeper into nature, we have to abandon more and more of the images and concepts of ordinary language.

On this journey to the world of the infinitely small, the most important step, from a philosophical point of view, was the first one: the step into the world of atoms. Probing inside the atom and investigating its structure, science transcended the limits of our sensory imagination. From this point on, it could no longer rely with absolute certainty on logic and common sense. Atomic physics provided the scientists with the first glimpses of the essential nature of things. Like the mystics, physicists were now dealing with a nonsensory experience of reality and, like the mystics, they had to face the paradoxical aspects of this experience. From then on therefore, the models and images of modern physics became akin to those of Eastern philosophy.

4
THE NEW PHYSICS

According to the Eastern mystics, the direct mystical experience of reality is a momentous event which shakes the very foundations of one's world view. D. T. Suzuki has called it "the most startling event that could ever happen in the realm of human consciousness . . . upsetting every form of standardized experience,"[1] and he has illustrated

the shocking character of this experience with the words of a Zen master who described it as "the bottom of a pail breaking through."

Physicists, at the beginning of this century, felt much the same way when the foundations of their world-view were shaken by the new experience of the atomic reality, and they described this experience in terms which were often very similar to those used by Suzuki's Zen master. Thus Heisenberg wrote:

> The violent reaction on the recent development of modern physics can only be understood when one realizes that here the foundations of physics have started moving; and that this motion has caused the feeling that the ground would be cut from science.[2]

Einstein experienced the same shock when he first came in contact with the new reality of atomic physics. He wrote in his autobiography:

> All my attempts to adapt the theoretical foundation of physics to this [new type of] knowledge failed completely. It was as if the ground had been pulled out from under one, with no firm foundation to be seen anywhere, upon which one could have built.[3]

The discoveries of modern physics necessitated profound changes of concepts like space, time, matter, object, cause and effect, etc.; and since these concepts are so basic to our way of experiencing the world, it is not surprising that the physicists who were forced to change them felt something of a shock. Out of these changes emerged a new and radically different world-view, still in the process of formation by current scientific research.

It seems, then, that Eastern mystics and Western physicists went through similar revolutionary experiences which led them to completely new ways of seeing the world. In the following two passages, the European physicist Niels Bohr and the Indian mystic Sri Aurobindo both express the depth and the radical character of this experience.

> The great extension of our experience in recent years has brought to light the insufficiency of our simple mechanical conceptions and, as a consequence, has shaken the foundation on which the customary interpretation of observation was based.[4]
>
> *Niels Bohr*

> All things in fact begin to change their nature and appearance; one's whole experience of the world is radically different. . . . There is a new vast and deep way of experiencing, seeing, knowing, contacting things.[5]
>
> *Sri Aurobindo*

This chapter will serve to sketch a preliminary picture of this new conception of the world against the contrasting background of classical physics;* showing how the

* The reader who finds this preliminary presentation of modern physics too compressed and difficult to understand should not be unduly worried. All of the concepts mentioned in this chapter will be discussed in greater detail later on.

classical mechanistic world-view had to be abandoned at the beginning of this century when quantum theory and relativity theory—the two basic theories of modern physics—forced us to adopt a much more subtle, holistic and "organic" view of nature.

CLASSICAL PHYSICS

The world-view which was changed by the discoveries of modern physics had been based on Newton's mechanical model of the universe. This model constituted the solid framework of classical physics. It was indeed a most formidable foundation supporting, like a mighty rock, all of science and providing a firm basis for natural philosophy for almost three centuries.

The stage of the Newtonian universe, on which all physical phenomena took place, was the three-dimensional space of classical Euclidean geometry. It was an absolute space, always at rest and unchangeable. In Newton's own words, "Absolute space, in its own nature, without regard to anything external, remains always similar and immovable."[6] All changes in the physical world were described in terms of a separate dimension, called time, which again was absolute, having no connection with the material world and flowing smoothly from the past through the present to the future. "Absolute, true, and mathematical time," said Newton, "of itself and by its own nature, flows uniformly, without regard to anything external."[7]

The elements of the Newtonian world which moved in this absolute space and absolute time were material particles. In the mathematical equations they were treated as "mass points," and Newton saw them as small, solid, and indestructible objects out of which all matter was made. This model was quite similar to that of the Greek atomists. Both were based on the distinction between the full and the void, between matter and space, and in both models the particles remained always identical in their mass and shape. Matter was therefore always conserved and essentially passive. The important difference between the Democritean and Newtonian atomism is that the latter includes a precise description of the force acting between the material particles. This force is very simple, depending only on the masses and the mutual distances of the particles. It is the force of gravity, and it was seen by Newton as rigidly connected with the bodies it acted upon, and as acting instantaneously over a distance. Although this was a strange hypothesis, it was not investigated further. The particles and the forces between them were seen as created by God and thus were not subject to further analysis. In his *Opticks*, Newton gives us a clear picture of how he imagined God's creation of the material world:

> It seems probable to me that God in the beginning formed matter in solid, massy, hard, impenetrable, movable particles, of such sizes and figures, and with such other properties, and in such proportion to space, as most conduced to the end for which he formed them; and that these primitive particles being solids, are incomparably harder than any porous bodies compounded of them; even so very hard, as never to wear or break in pieces; no ordinary power being able to divide what God himself made one in the first creation.[8]

All physical events are reduced, in Newtonian mechanics, to the motion of material points in space, caused by their mutual attraction, i.e., by the force of gravity. In order to put the effect of this force on a mass point into a precise mathematical form, Newton had to invent completely new concepts and mathematical techniques, those of differential calculus. This was a tremendous intellectual achievement and has been praised by Einstein as "perhaps the greatest advance in thought that a single individual was ever privileged to make."

Newton's equations of motion are the basis of classical mechanics. They were considered to be fixed laws according to which material points move, and were thus thought to account for all changes observed in the physical world. In the Newtonian view, God had created, in the beginning, the material particles, the forces between them, and the fundamental laws of motion. In this way, the whole universe was set in motion and it has continued to run ever since, like a machine, governed by immutable laws.

The mechanistic view of nature is thus closely related to a rigorous determinism. The giant cosmic machine was seen as being completely causal and determinate. All that happened had a definite cause and gave rise to a definite effect, and the future of any part of the system could—in principle—be predicted with absolute certainty if its state at any time was known in all details. This belief found its clearest expression in the famous words of the French mathematician Pierre Simon Laplace:

> An intellect which at a given instant knew all the forces acting in nature, and the position of all things of which the world consists—supposing the said intellect were vast enough to subject these data to analysis—would embrace in the same formula the motions of the greatest bodies in the universe and those of the slightest atoms; nothing would be uncertain for it, and the future, like the past, would be present to its eyes.[9]

The philosophical basis of this rigorous determinism was the fundamental division between the I and the world introduced by Descartes. As a consequence of this division, it was believed that the world could be described objectively, i.e., without ever mentioning the human observer, and such an objective description of nature became the ideal of all science.

The eighteenth and nineteenth centuries witnessed a tremendous success of Newtonian mechanics. Newton himself applied his theory to the movement of the planets and was able to explain the basic features of the solar system. His planetary model was greatly simplified, however, neglecting, for example, the gravitational influence of the planets on each other, and thus he found that there were certain irregularities which he could not explain. He resolved this problem by assuming that God was always present in the universe to correct these irregularities.

Laplace, the great mathematician, set himself the ambitious task of refining and perfecting Newton's calculations in a book which should "offer a complete solution of the great mechanical problem presented by the solar system, and bring theory to coincide so closely with observation that empirical equations would no longer find a place in astronomical tables."[10] The result was a large work in five volumes, called *Mécanique Céleste* in which Laplace succeeded in explaining the motions of the planets, moons and comets down to the smallest details, as well as the flow of the tides and other phenomena related to gravity. He showed that the Newtonian laws of motion assured the stability of the solar system and treated the universe as a perfectly self-regulating machine. When Laplace presented the first edition of his work to Napoleon—so the story goes—Napoleon remarked, "Monsieur Laplace, they tell me you have written this large book on the system of the universe, and have never even mentioned its Creator." To this Laplace replied bluntly, "I had no need for that hypothesis."

Encouraged by the brilliant success of Newtonian mechanics in astronomy, physicists extended it to the continuous motion of fluids and to the vibrations of elastic bodies, and again it worked. Finally, even the theory of heat could be reduced to mechanics when it was realized that heat was the energy created by a complicated "jiggling" motion of the molecules. When the temperature of, say, water is increased the motion of the water molecules increases until they overcome the forces holding them together and fly apart. In this way, water turns into steam. On the other hand, when the thermal motion is slowed down by cooling the water, the molecules finally lock into a new, more rigid pattern which is ice. In a similar way, many other thermal phenomena can be understood quite well from a purely mechanistic point of view.

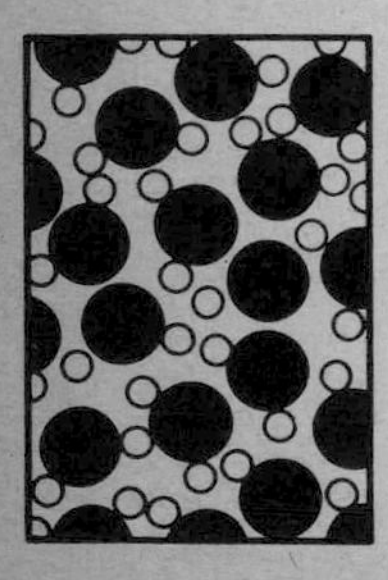

water

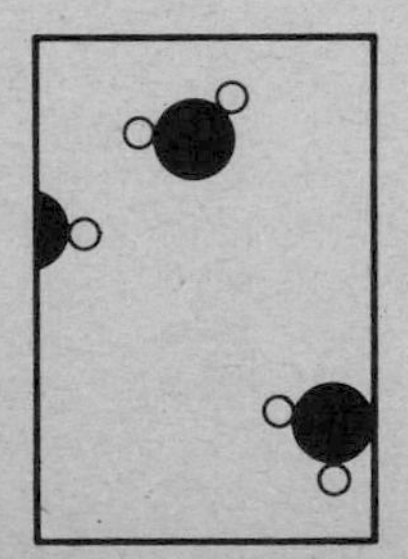

steam

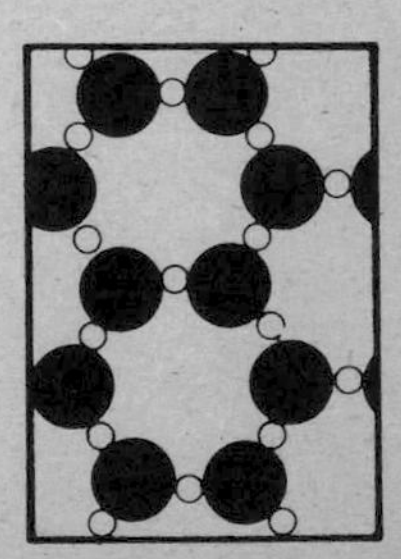

ice

The enormous success of the mechanistic model made physicists of the early nineteenth century believe that the universe was indeed a huge mechanical system running according to the Newtonian laws of motion. These laws were seen as the basic laws of nature, and Newton's mechanics were considered to be the ultimate theory of natural phenomena. And yet, it was less than a hundred years later that a new physical reality was discovered which made the limitations of the Newtonian model apparent and showed that none of its features had absolute validity.

This realization did not come abruptly, but was initiated by developments that had already started in the nineteenth century and prepared the way for the scientific revolutions of our time. The first of these developments was the discovery and investigation of electric and magnetic phenomena which could not be described appropriately by the mechanistic model and involved a new type of force. The important step was made by Michael Faraday and Clerk Maxwell—the first, one of the greatest experimenters in the history of science, the second, a brilliant theorist. When Faraday produced an electric current in a coil of copper by moving a magnet near it, and thus converted the mechanical work of moving the magnet into electric energy, he brought science and technology to a turning point. His fundamental experiment gave birth, on the one hand, to the vast technology of electrical engineering; on the other hand, it formed the basis of his and Maxwell's theoretical speculations which, eventually, resulted in a complete theory of electromagnetism. Faraday and Maxwell not only studied the effects of the electric and magnetic forces, but made the forces themselves the primary object of their investigation. They replaced the concept of a force by that of a force field, and in doing so they were the first to go beyond Newtonian physics.

Instead of interpreting the interaction between a positive and a negative charge simply by saying that the two charges attract each other like two masses in Newtonian mechanics, Faraday and Maxwell found it more appropriate to say that each charge creates a "disturbance," or a "condition," in the space around it so that the other charge, when it is present, feels a force.

This condition in space which has the potential of producing a force is called a field. It is created by a single charge and it exists whether or not another charge is brought in to feel its effect.

This was a most profound change in man's conception of physical reality. In the Newtonian view, the forces were rigidly connected with the bodies they act upon. Now the force concept was replaced by the much subtler concept of a field which had its own reality and could be studied without any reference to material bodies. The culmination of this theory, called electrodynamics, was the realization that light is nothing but a rapidly alternating electromagnetic field traveling through space in the form of waves. Today we know that radio waves, light waves, and X rays are all electromagnetic waves, oscillating electric and magnetic fields differing only in the frequency of their oscillation, and that visible light is only a tiny fraction of the electromagnetic spectrum.

In spite of these far-reaching changes, Newtonian mechanics at first held its position as the basis of all physics. Maxwell himself tried to explain his results in mechanical terms, interpreting the fields as states of mechanical stress in a very light space-filling medium, called ether, and the electromagnetic waves as elastic waves of this ether. This was only natural, as waves are usually experienced as vibrations of something; water waves as vibrations of water, sound waves as vibrations of air. Maxwell, however, used several mechanical interpretations of his theory at the same time and apparently took none of them really seriously. He must have realized intuitively, even if he did not say so explicitly, that the fundamental entities in his theory were the fields and not the mechanical models. It was Einstein who clearly recognized this fact fifty years later when he declared that no ether existed and that the electromagnetic fields were physical entities in their own right which could travel through empty space and could not be explained mechanically.

At the beginning of the twentieth century, then, physicists had two successful theories which applied to different phenomena: Newton's mechanics and Maxwell's electrodynamics. Thus the Newtonian model had ceased to be the basis of all physics.

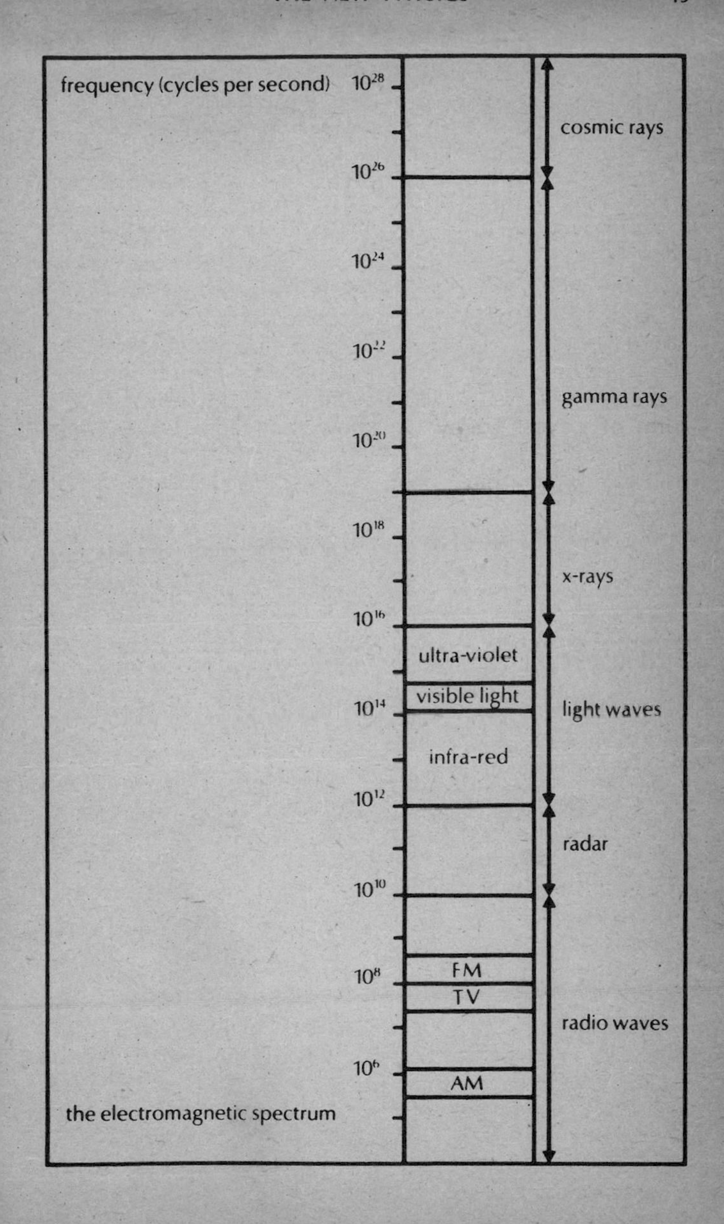
frequency (cycles per second)
10^28
10^26
10^24
10^22
10^20
10^18
10^16
10^14
10^12
10^10
10^8
10^6
cosmic rays
gamma rays
x-rays
ultra-violet
visible light
infra-red
light waves
radar
FM
TV
AM
radio waves
the electromagnetic spectrum

MODERN PHYSICS

The first three decades of our century changed the whole situation in physics radically. Two separate developments —that of relativity theory and of atomic physics— shattered all the principal concepts of the Newtonian world view: the notion of absolute space and time, the elementary solid particles, the strictly causal nature of physical phenomena, and the ideal of an objective description of nature. None of these concepts could be extended to the new domains into which physics was now penetrating.

At the beginning of modern physics stands the extraordinary intellectual feat of one man: Albert Einstein. In two articles, both published in 1905, Einstein initiated two revolutionary trends of thought. One was his special theory of relativity; the other was a new way of looking at electromagnetic radiation which was to become characteristic of quantum theory, the theory of atomic phenomena. The complete quantum theory was worked out twenty years later by a whole team of physicists. Relativity theory, however, was constructed in its complete form almost entirely by Einstein himself. Einstein's scientific papers stand at the beginning of the twentieth century as imposing intellectual monuments—the pyramids of modern civilization.

Einstein strongly believed in nature's inherent harmony, and his deepest concern throughout his scientific life was to find a unified foundation of physics. He began to move toward this goal by constructing a common framework for electrodynamics and mechanics, the two separate theories of classical physics. This framework is known as the special theory of relativity. It unified and completed the structure of classical physics, but at the same time it involved drastic changes in the traditional concepts of space and time and undermined one of the foundations of the Newtonian world view.

According to relativity theory, space is not three-dimensional and time is not a separate entity. Both are intimately connected and form a four-dimensional continuum, "space-time." In relativity theory, therefore, we can never talk about space without talking about time and vice versa. Furthermore, there is no universal flow of time, as in the Newtonian model. Different observers will order events differently in time if they move with different velocities relative to the observed events. In such a case, two events which are seen as occurring simultaneously by one observer may occur in different temporal sequences for other observers. All measurements involving space and time thus lose their absolute significance. In relativity theory, the Newtonian concept of an absolute space as the stage of physical phenomena is abandoned, and so is the concept of an absolute time. Both space and time become merely elements of the language a particular observer uses for his description of the phenomena.

The concepts of space and time are so basic for the description of natural phenomena that their modification entails a modification of the whole framework that we use to describe nature. The most important consequence of this modification is the realization that mass is nothing but a form of energy. Even an object at rest has energy stored in its mass, and the relation between the two is given by the famous equation $E = mc^2$, c being the speed of light.

This constant c, the speed of light, is of fundamental importance for the theory of relativity. Whenever we describe physical phenomena involving velocities which approach the speed of light, our description has to take relativity theory into account. This applies in particular to electromagnetic phenomena, of which light is just one example and which led Einstein to the formulation of his theory.

In 1915 Einstein proposed his general theory of relativity in which the framework of the special theory is extended to include gravity, i.e., the mutual attraction of all massive bodies. Whereas the special theory has been confirmed by innumerable experiments, the general theory has not yet been confirmed conclusively. However, it is so far the most accepted, consistent and elegant theory of gravity and is widely used in astrophysics and cosmology for the description of the universe at large.

The force of gravity, according to Einstein's theory, has the effect of "curving" space and time. This means that ordinary Euclidean geometry is no longer valid in such a curved space, just as the two-dimensional geometry of a plane cannot be applied on the surface of a sphere. On a plane, we can draw, for example, a square by marking off one meter on a straight line, making a right angle and marking off another meter, then making another right angle and marking off another meter, and finally making a third right angle and marking off one meter again, after which we are back at the starting point and the square is completed. On a sphere, however, this procedure does not work because the rules of Euclidean geometry do not hold on curved surfaces. In the same way, we can define a three-dimensional curved space to be one in which Euclidean geometry is no longer valid. Einstein's theory, now, says that three-dimensional space is actually curved, and that the curvature is caused by the gravitational field of massive bodies.

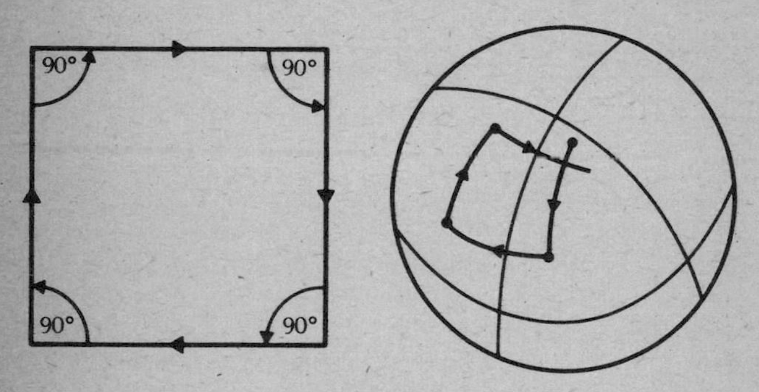

drawing a square on a plane and on a sphere

Wherever there is a massive object, e.g., a star or a planet, the space around it is curved and the degree of curvature depends on the mass of the object. And as space can never be separated from time in relativity theory, time as well is affected by the presence of matter, flowing at different rates in different parts of the universe. Einstein's general theory of relativity thus completely abolishes the concepts of absolute space and time. Not only are all measurements involving space and time relative; the whole structure of space-time depends on the distribution of matter in the universe, and the concept of "empty space" loses its meaning.

The mechanistic world view of classical physics was based on the notion of solid bodies moving in empty space. This notion is still valid in the region that has been called the "zone of middle dimensions," that is, in the realm of our daily experience where classical physics continues to be a useful theory. Both concepts—that of empty space and that of solid material bodies—are deeply ingrained in our habits of thought, so it is extremely difficult for us to imagine a physical reality where they do not apply. And yet, this is precisely what modern physics forces us to do when we go beyond the middle dimensions. "Empty space" has lost its meaning in astrophysics and cosmology, the sciences of the universe at large, and the concept of solid objects was shattered by atomic physics, the science of the infinitely small.

At the turn of the century, several phenomena connected with the structure of atoms and inexplicable in terms of classical physics were discovered. The first indication that atoms had some structure came from the discovery of X rays; a new radiation which rapidly found its now-well-known application in medicine. X rays, however, are not the only radiation emitted by atoms. Soon after their discovery, other kinds of radiation were discovered which are emitted by the atoms of so-called radioactive substances. The phenomenon of radioactivity gave definite proof of the composite nature of atoms, showing that the atoms of radioactive substances not only emit various types of radiation, but also transform themselves into atoms of completely different substances.

Besides being objects of intense study, these phenomena were also used, in most ingenious ways, as new tools to probe deeper into matter than had ever been possible before. Thus Max von Laue used X rays to study the arrangements of atoms in crystals, and Ernest Rutherford realized that the so-called alpha particles emanating from radioactive substances were high-speed projectiles of subatomic size which could be used to explore the interior of the atom. They could be fired at atoms, and from the way they were deflected one could draw conclusions about the atoms' structure.

When Rutherford bombarded atoms with these alpha particles, he obtained sensational and totally unexpected results. Far from being the hard and solid particles they were believed to be since antiquity, the atoms turned out to consist of vast regions of space in which extremely small particles—the electrons—moved around the nucleus, bound to it by electric forces. It is not easy to get a feeling for the order of magnitude of atoms, so far is it removed from our macroscopic scale. The diameter of an atom is about one hundred-millionth of a centimeter. In order to visualize this diminutive size, imagine an orange blown up to the size of the earth. The atoms of the orange will then have the size of cherries. Myriads of cherries, tightly packed into a globe of the size of the earth—that's a magnified picture of the atoms in an orange.

An atom, therefore, is extremely small compared to macroscopic objects, but it is huge compared to the nucleus in its center. In our picture of cherry-sized atoms, the nucleus of an atom will be so small that we will not be able to see it. If we blew up the atom to the size of a football, or even to room size, the nucleus would still be too small to be seen by the naked eye. To see the nucleus, we would have to blow up the atom to the size of the biggest dome in the world, the dome of St. Peter's Cathedral in Rome. In an atom of that size, the nucleus would have the size of a grain of salt! A grain of salt in the middle of the dome of St. Peter's, and specks of dust whirling around it in the vast space of the dome—this is how we can picture the nucleus and electrons of an atom.

Soon after the emergence of this "planetary" model of the atom, it was discovered that the number of electrons in the atoms of an element determine the element's chemical properties, and today we know that the whole periodic table of elements can be built up by successively adding protons and neutrons to the nucleus of the lightest atom—hydrogen*—and the corresponding number of electrons to its atomic "shell." The interactions between the atoms give rise to the various chemical processes, so that all of chemistry can now in principle be understood on the basis of the laws of atomic physics.

* The hydrogen atom consists of just one proton and one electron.

These laws, however, were not easy to recognize. They were discovered in the 1920s by an international group of physicists including Niels Bohr from Denmark, Louis de Broglie from France, Erwin Schrödinger and Wolfgang Pauli from Austria, Werner Heisenberg from Germany, and Paul Dirac from England. These men joined their forces across all national borders and shaped one of the most exciting periods in modern science, which brought man, for the first time, into contact with the strange and unexpected reality of the subatomic world. Every time the physicists asked nature a question in an atomic experiment, nature answered with a paradox, and the more they tried to clarify the situation, the sharper the paradoxes became. It took them a long time to accept the fact that these paradoxes belong to the intrinsic structure of atomic physics, and to realize that they arise whenever one attempts to describe atomic events in the traditional terms of physics. Once this was perceived, the physicists began to learn to ask the right questions and to avoid contradictions. In the words of Heisenberg, "They somehow got into the spirit of the quantum theory," and finally they found the precise and consistent mathematical formulation of this theory.

The concepts of quantum theory were not easy to accept even after their mathematical formulation had been completed. Their effect on the physicists' imaginations was truly shattering. Rutherford's experiments had shown that atoms, instead of being hard and indestructible, consisted of vast regions of space in which extremely small particles moved, and now quantum theory made it clear that even these particles were nothing like the solid objects of classical physics. The subatomic units of matter are very abstract entities which have a dual aspect. Depending on how we look at them, they appear sometimes as particles, sometimes as waves; and this dual nature is also exhibited by light which can take the form of electromagnetic waves or of particles.

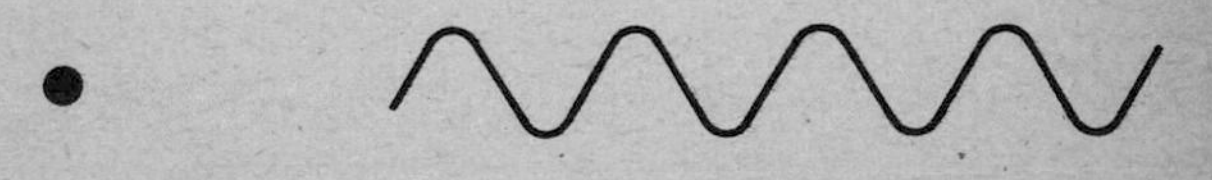

a particle a wave

This property of matter and of light is very strange. It seems impossible to accept that something can be, at the same time, a particle—i.e., an entity confined to a very small volume—and a wave, which is spread out over a large region of space. This contradiction gave rise to most of the *koan*-like paradoxes which finally led to the formulation of quantum theory. The whole development started when Max Planck discovered that the energy of heat radiation is not emitted continuously, but appears in the form of "energy packets." Einstein called these energy packets "quanta" and recognized them as a fundamental aspect of nature. He was bold enough to postulate that light and every other form of electromagnetic radiation can appear not only as electromagnetic waves, but also in the form of these quanta. The light quanta, which gave quantum theory its name, have since been accepted as bona fide particles and are now called photons. They are particles of a special kind, however, massless and always traveling with the speed of light.

The apparent contradiction between the particle and the wave picture was solved in a completely unexpected way which called in question the very foundation of the mechanistic world-view—the concept of the reality of matter. At the subatomic level, matter does not exist with certainty at definite places, but rather shows "tendencies to exist," and atomic events do not occur with certainty at definite times and in definite ways, but rather show "tendencies to occur." In the formalism of quantum theory, these tendencies are expressed as probabilities and are associated with mathematical quantities which take the form of waves. This is why particles can be waves at the same time. They are not "real" three-dimensional waves like sound or water waves. They are "probability waves," abstract mathematical quantities with all the characteristic properties of waves which are related to the probabilities of finding the particles at particular points in space and at particular times. All the laws of atomic physics are expressed in terms of these probabilities. We can never predict an atomic event with certainty; we can say only how likely it is to happen.

Quantum theory has thus demolished the classical concepts of solid objects and of strictly deterministic laws of nature. At the subatomic level, the solid material objects of classical physics dissolve into wavelike patterns of probabilities, and these patterns, ultimately, do not represent probabilities of things, but rather probabilities of interconnections. A careful analysis of the process of observation in atomic physics has shown that the subatomic particles have no meaning as isolated entities, but can only be understood as interconnections between the preparation of an experiment and the subsequent measurement. Quantum theory thus reveals a basic oneness of the universe. It shows that we cannot decompose the world into independently existing smallest units. As we penetrate into matter, nature does not show us any isolated "basic building blocks," but rather appears as a complicated web of relations between the various parts of the whole. These relations always include the observer in an essential way. The human observer constitutes the final link in the chain of observational processes, and the properties of any atomic object can be understood only in terms of the object's interaction with the observer. This means that the classical ideal of an objective description of nature is no longer valid. The Cartesian partition between the I and the world, between the observer and the observed, cannot be made when dealing with atomic matter. In atomic physics, we can never speak about nature without, at the same time, speaking about ourselves.

The new atomic theory could immediately solve several puzzles which had arisen in connection with the structure of atoms and could not be explained by Rutherford's planetary model. First of all, Rutherford's experiments had shown that the atoms making up solid matter consist almost entirely of empty space, as far as the distribution of mass is concerned. But if all the objects around us, and we ourselves, consist mostly of empty space, why can't we walk through closed doors? In other words, what is it that gives matter its solid aspect?

A second puzzle was the extraordinary mechanical stability of atoms. In the air, for example, atoms collide millions of times every second and yet return to their original form after each collision. No planetary system following the laws of classical mechanics would ever come out of these collisions unaltered. But an oxygen atom will always retain its characteristic configuration of electrons, no matter how often it collides with other atoms. This configuration, furthermore, is exactly the same in all atoms of a given kind. Two iron atoms, and consequently two pieces of pure iron, are completely identical, no matter where they come from or how they have been treated in the past.

Quantum theory has shown that all these astonishing properties of atoms arise from the wave nature of their electrons. To begin with, the solid aspect of matter is the consequence of a typical "quantum effect" connected with the dual wave/particle aspect of matter, a feature of the subatomic world which has no macroscopic analogue. Whenever a particle is confined to a small region of space, it reacts to this confinement by moving around, and the smaller the region of confinement is, the faster the particle moves around in it. In the atom, now, there are two competing forces. On the one hand, the electrons are bound to the nucleus by electric forces which try to keep them as close as possible. On the other hand, they respond to their confinement by whirling around, and the tighter they are bound to the nucleus, the higher their velocity will be; in fact, the confinement of electrons in an atom results in enormous velocities of about 600 miles per second! These high velocities make the atom appear as a rigid sphere, just as a fast-rotating propeller appears as a disc. It is very difficult to compress atoms any further, and thus they give matter its familiar solid aspect.

In the atom, then, the electrons settle in orbits in such a way that there is an optimal balance between the attraction of the nucleus and their reluctance to be confined. The atomic orbits, however, are very different from those of the planets in the solar system, the difference arising from the wave nature of the electrons. An atom cannot be pictured as a small planetary system. Rather than particles circling around the nucleus, we have to imagine probability waves arranged in different orbits. Whenever we make a measurement, we will find the electrons somewhere in these orbits, but we cannot say that they are "going around the nucleus" in the sense of classical mechanics.

In the orbits, the electron waves have to be arranged in such a way that "their ends meet," i.e., that they form patterns known as "standing waves." These patterns appear whenever waves are confined to a finite region, like the waves in a vibrating guitar string, or in the air inside a flute (see diagram below). It is well known from

standing-wave patterns in a vibrating string

these examples that standing waves can assume only a limited number of well-defined shapes. In the case of the electron waves inside an atom, this means that they can exist only in certain atomic orbits with definite diameters. The electron of a hydrogen atom, for example, can exist only in a certain first, second or third orbit, etc., and nowhere in between. Under normal conditions, it will always be in its lowest orbit, called the "ground state" of the atom. From there, the electron can jump to higher orbits if it receives the necessary amount of energy, and then the atom is said to be in an "excited state" from which it will go back to its ground state after a while, the electron giving off the surplus energy in the form of a quantum of electromagnetic radiation, or photon. The states of an atom—i.e., the shapes and mutual distances of its electron orbits—are exactly the same for all atoms with the same number of electrons. This is why any two oxygen atoms, for example, will be completely identical. They may be in different excited states, perhaps due to collisions with other atoms in the air, but after a while they will invariably return to exactly the same ground state. The wave nature of the electrons accounts thus for the identity of atoms and for their great mechanical stability.

A further characteristic feature of atomic states is the fact that they can be completely specified by a set of integral numbers, called "quantum numbers," which indicate the location and shape of the electron orbits. The first quantum number is the number of the orbit and determines the energy an electron must have to be in that orbit; two more numbers specify the detailed shape of the electron wave in the orbit and are related to the speed and orientation of the electron's rotation.* The fact that these details are expressed by integral numbers means that the electron cannot change its rotation continuously, but can only jump from one value to another, just as it can only jump from one orbit to another. Again the higher values represent excited states of the atom, the ground state being the one where all the electrons are in the lowest possible orbits and have the smallest possible amounts of rotation.

Tendencies to exist, particles reacting to confinement with motion, atoms switching suddenly from one "quantum state" to another, and an essential interconnectedness of all phenomena—these are some of the unusual features of the atomic world. The basic force, on the other hand, which gives rise to all atomic phenomena is

* The "rotation" of an electron in its orbit must not be understood in the classical sense; it is determined by the shape of the electron wave in terms of the probabilities for the particle's existence in certain parts of the orbit.

familiar and can be experienced in the macroscopic world. It is the force of electric attraction between the positively charged atomic nucleus and the negatively charged electrons. The interplay of this force with the electron waves gives rise to the tremendous variety of structures and phenomena in our environment. It is responsible for all chemical reactions, and for the formation of molecules, that is, of aggregates of several atoms bound to each other by mutual attraction. The interaction between electrons and atomic nuclei is thus the basis of all solids, liquids, and gases, and also of all living organisms and of the biological processes associated with them.

In this immensely rich world of atomic phenomena, the nuclei play the role of extremely small, stable centers which constitute the source of the electric force and form the skeletons of the great variety of molecular structures. To understand these structures, and most of the natural phenomena around us, it is not necessary to know more about the nuclei than their charge and their mass. In order to understand the nature of matter, however, to know what matter is ultimately made of, one has to study the atomic nuclei which contain practically all of its mass. In the 1930s, after quantum theory had unraveled the world of atoms, it was therefore the main task of physicists to understand the structure of nuclei, their constituents and the forces which hold them together so tightly.

The first important step towards an understanding of nuclear structure was the discovery of the neutron as the second constituent of the nucleus, a particle which has roughly the same mass as the proton (the first nuclear constituent)—about two thousand times the mass of the electron—but does not carry an electric charge. This discovery not only explained how the nuclei of all chemical elements were built up from protons and neutrons, but also revealed that the nuclear force, which kept these particles so tightly bound within the nucleus, was a completely new phenomenon. It could not be of electromagnetic origin since the neutrons were electrically neutral. Physicists soon realized that they were here confronted with a new force of nature which does not manifest itself anywhere outside the nucleus.

An atomic nucleus is about one hundred thousand times smaller than the whole atom, and yet it contains almost all of the atom's mass. This means that matter inside the nucleus must be extremely dense compared to the forms of matter we are used to. Indeed, if the whole human body were compressed to nuclear density, it would not take up more space than a pinhead. This high density, however, is not the only unusual property of nuclear matter. Being of the same quantum nature as electrons, the "nucleons"—as the protons and neutrons are often called—respond to their confinement with high velocities, and since they are squeezed into a much smaller volume their reaction is all the more violent. They race about in the nucleus with velocities of about 40,000 miles per second! Nuclear matter is thus a form of matter entirely different from anything we experience "up here" in our macroscopic environment. We can, perhaps, picture it best as tiny drops of an extremely dense liquid which is boiling and bubbling most fiercely.

The essential new aspect of nuclear matter which accounts for all its unusual properties is the strong nuclear force, and the feature that makes this force so unique is its extremely short range. It acts only when the nucleons come very near to each other, that is, when their distance is about two to three times their diameter. At such a distance, the nuclear force is strongly attractive; but when the distance becomes less, the force becomes strongly repulsive so that the nucleons cannot approach each other any closer. In this way, the nuclear force keeps the nucleus in an extremely stable, though extremely dynamic equilibrium.

The picture of matter which emerges from the study of atoms and nuclei shows that most of it is concentrated in tiny drops separated by huge distances. In the vast space between the massive and fiercely boiling nuclear drops move the electrons. These constitute only a tiny fraction of the total mass, but give matter its solid aspect and provide the links necessary to build up the molecular structures. They are also involved in the chemical reactions and are responsible for the chemical properties of matter. Nuclear reactions, on the other hand, generally do not occur naturally in this form of matter because the available energies are not high enough to disturb the nuclear equilibrium.

This form of matter, however, with its multitude of shapes and textures and its complicated molecular architecture, can exist only under very special conditions, when the temperature is not too high, so that the molecules do not jiggle too much. When the thermal energy increases about a hundredfold, as it does in most stars, all atomic and molecular structures are destroyed. Most of the matter in the universe exists, in fact, in a state which is very different from the one just described. In the center of the stars exist large accumulations of nuclear matter, and nuclear processes which occur only very rarely on earth predominate there. They are essential for the great variety of stellar phenomena observed in astronomy, most of which arise from a combination of nuclear and gravitational effects. For our planet, the nuclear processes in the center of the sun are of particular importance because they furnish the energy which sustains our terrestrial environment. It has been one of the great triumphs of modern physics to discover that the constant energy flow from the sun, our vital link with the world of the very large, is a result of nuclear reactions, of phenomena in the world of the Infinitely small.

In the history of man's penetration into this submicroscopic world, a stage was reached in the early 1930s when scientists thought they had now finally discovered the "basic building blocks" of matter. It was known that all matter consisted of atoms and that all atoms consisted of protons, neutrons, and electrons. These co-called "elementary particles" were seen as the ultimate indestructible units of matter: atoms in the Democritean sense. Although quantum theory implies, as mentioned previously, that we cannot decompose the world into independently existing smallest units, this was not generally perceived at that time. The classical habits of thought were still so persistent that most physicists tried to understand matter in terms of its "basic building blocks," this trend of thought is, in fact, quite strong even today.

Two further developments in modern physics have

shown, however, that the notion of elementary particles as the primary units of matter has to be abandoned. One of these developments was experimental, the other theoretical, and both began in the 1930s. On the experimental side, new particles were discovered as physicists

Meson Table

April 1974

entry	$I^G(J^P)C_n$	entry	$I^G(J^P)C_n$	entry	$I^G(J^P)C_n$	entry	$I(J^P)$
π (140)	$1^-(0^-)+$	→ η_N (1080)	$0^+(N)^+$	ρ' (1600)	$1^+(1^-)-$	K (494)	$1/2(0^-)$
η (549)	$0^+(0^-)^+$	A_1 (1100)	$1^-(1^+)+$	A_3 (1640)	$1^-(2^-)+$	K* (892)	$1/2(1^-)$
ϵ (600)	$0^+(0^+)+$	→ M (1150)		ω (1675)	$0^-(N)-$	κ	$1/2(0^+)$
ρ (770)	$1^+(1^-)-$	→ $A_{1.5}$ (1170)	1^-	g (1680)	$1^+(3^-)-$	Q	$1/2(1^+)$
ω (783)	$0^-(1^-)-$	B (1235)	$1^+(1^+)-$	→ X (1690)	$-$	K* (1420)	$1/2(2^+)$
→ M (940)		→ ρ' (1250)	$1^+(1^-)-$	→ X (1795)	1	→ K_N (1660)	1/2
→ M (953)	$+$	f (1270)	$0^+(2^+)+$	→ S (1930)	1	→ K_N (1760)	1/2
η' (958)	$0^+(0^-)+$	D (1285)	$0^+(A)+$	→ A_4 (1960)	1^-	L (1770)	$1/2(A)$
δ (970)	$1^-(0^+)+$	A_2 (1310)	$1^-(2^+)$	→ ρ (2100)	1^+	→ K_N (1850)	
→ H (990)	$0^-(A)-$	E (1420)	$0^+(A)+$	→ T (2200)	1	→ K* (2200)	
S* (993)	$0^+(0^+)+$	→ X (1430)	0	→ ρ (2275)	1^+	→ K* (2800)	
ϕ (1019)	$0^-(1^-)-$	→ X (1440)	1	→ U (2360)	1		
→ M (1033)		f′ (1514)	$0^+(2^+)+$	→ $N\bar{N}$ (2375)	0	→ Exotics	
→ B_1 (1040)	1^+	F_1 (1540)	$1\ (A)$	→ X(2500–3600)			

refined their experimental techniques and developed ingenious new devices for particle detection. Thus the number of particles increased from three to six by 1935, then to eighteen by 1955, and today we know over two hundred "elementary" particles. The two tables taken from a recent publication,[11] show most of the particles known today. They illustrate convincingly that the adjective "elementary" is no longer very attractive in such a situation. As more and more particles were discovered over the years, it became clear that not all of them could be called "elementary," and today there is a widespread

Baryon Table

April 1974

N(939)	P11	****	Δ(1232)	P33	****	Λ(1116)	P01	****	Σ(1193)	P11	****	Ξ(1317)	P11	****
N(1470)	P11	****	Δ(1650)	S31	****	Λ(1330)		Dead	Σ(1385)	P13	****	Ξ(1530)	P13	****
N(1520)	D13	****	Δ(1670)	D33	***	Λ(1405)	S01	****	Σ(1440)		Dead	Ξ(1630)		**
N(1535)	S11	****	Δ(1690)	P33	*	Λ(1520)	D03	****	Σ(1480)		*	Ξ(1820)		***
N(1670)	D15	****	Δ(1890)	F35	***	Λ(1670)	S01	****	Σ(1620)	S11	**	Ξ(1940)		***
N(1688)	F15	****	Δ(1900)	S31	*	Λ(1690)	D03	****	Σ(1620)	P11	**	Ξ(2030)		**
N(1700)	S11	****	Δ(1910)	P31	***	Λ(1750)	P01	**	Σ(1670)	D13	****	Ξ(2250)		*
N(1700)	D13	**	Δ(1950)	F37	****	Λ(1815)	F05	****	Σ(1670)		**	Ξ(2500)		**
N(1780)	P11	***	Δ(1960)	D35	**	Λ(1830)	D05	***	Σ(1690)		**			
N(1810)	P13	***	Δ(2160)		**	Λ(1860)	P03	**	Σ(1750)	S11	***	Ω(1672)	P03	****
N(1990)	F17	**	Δ(2420)	H311	***	Λ(1870)	S01	**	Σ(1765)	D15	****			
N(2000)	F15	**	Δ(2850)		***	Λ(2010)	D03	**	Σ(1840)	P13	*			
N(2040)	D13	**	Δ(3230)		***	Λ(2020)	F07	**	Σ(1880)	P11	**			
N(2100)	S11	*				Λ(2100)	G07	****	Σ(1915)	F15	****			
N(2100)	D15	*	Z0(1780)	P01	*	Λ(2110)	?05	*	Σ(1940)	D13	***			
N(2190)	G17	***	Z0(1865)	D03	*	Λ(2350)		****	Σ(2000)	S11	*			
N(2220)	H19	***	Z1(1900)	P13	*	Λ(2585)		***	Σ(2030)	F17	****			
N(2650)		***	Z1(2150)		*				Σ(2070)	F15	*			
N(3030)		***	Z1(2500)		*				Σ(2080)	P13	**			
N(3245)		*							Σ(2100)	G17	**			
N(3690)		*							Σ(2250)		****			
N(3755)		*							Σ(2455)		***			
									Σ(2620)		***			
									Σ(3000)		**			

**** Good, clear, and unmistakable. *** Good, but in need of clarification or not absolutely certain.
** Needs confirmation. * Weak.

belief among physicists that none of them deserves this name.

This belief is enforced by the theoretical developments which paralleled the discovery of an ever-increasing number of particles. Soon after the formulation of quantum theory, it became clear that a complete theory of nuclear phenomena must not only be a quantum theory, but must also incorporate relativity theory. This is because the particles confined to dimensions of the size of nuclei often move so fast that their speed comes close to the speed of light. This fact is crucial for the description of their behavior, because every description of natural phenomena involving velocities close to the speed of light has to take relativity theory into account. It has to be, as we say, a "relativistic" description. What we need, therefore, for a full understanding of the nuclear world is a theory which incorporates both quantum theory and relativity theory. Such a theory has not yet been found, and therefore we have as yet been unable to formulate a complete theory of the nucleus. Although we know quite a lot about nuclear structure and about the interactions between nuclear particles, we do not yet understand the nature and complicated form of the nuclear force on a fundamental level. There is no complete theory of the nuclear particle world comparable to quantum theory for the atomic world. We do have several "quantum-relativistic" models which describe some aspects of the world of particles very well, but the fusion of quantum and relativity theory into a complete theory of the particle world is still the central problem and great challenge of modern fundamental physics.

Relativity theory has had a profound influence on our picture of matter by forcing us to modify our concept of a particle in an essential way. In classical physics, the mass of an object had always been associated with an indestructible material substance, with some "stuff" of which all things were thought to be made. Relativity theory showed that mass has nothing to do with any substance, but is a form of energy. Energy, however, is a dynamic quantity associated with activity, or with processes. The fact that the mass of a particle is equivalent to a certain amount of energy means that the particle can no longer be seen as a static object, but has to be conceived as a dynamic pattern, a process involving the energy which manifests itself as the particle's mass.

This new view of particles was initiated by Dirac when he formulated a relativistic equation describing the behavior of electrons. Dirac's theory was not only extremely successful in accounting for the fine details of atomic structure, but also revealed a fundamental symmetry between matter and antimatter. It predicted the existence of an antielectron with the same mass as the electron, but with an opposite charge. This positively charged particle, now called the positron, was indeed discovered two years after Dirac had predicted it. The symmetry between matter and antimatter implies that for every particle there exists an antiparticle with equal mass and opposite charge. Pairs of particles and antiparticles can be created if enough energy is available and can be made to turn into pure energy in the reverse process of annihilation. These processes of particle creation and annihilation had been predicted from Dirac's theory before they were actually discovered in nature, and since then they have been observed millions of times.

The creation of material particles from pure energy is certainly the most spectacular effect of relativity theory, and it can be understood only in terms of the view of particles outlined above. Before relativistic particle physics, the constituents of matter had always been considered as being either elementary units which were indestructible and unchangeable, or as composite objects which could be broken up into their constituent parts; and the basic question was whether one could divide matter again and again, or whether one would finally arrive at some smallest indivisible units. After Dirac's discovery, the whole question of the division of matter appeared in a new light. When two particles collide with high energies, they generally break into pieces, but these pieces are not smaller than the original particles. They are again particles of the same kind and are created out of the energy of motion ("kinetic energy") involved in the collision process. The whole problem of dividing matter is thus resolved in an unexpected sense. The only way to divide subatomic particles further is to bang them together in collision processes involving high energies. This way, we can divide matter again and again, but we never obtain smaller pieces because we just create particles out of the energy involved in the process. The subatomic particles are thus destructible and indestructible at the same time.

This state of affairs is bound to remain paradoxical as long as we adopt the static view of composite "objects" consisting of "basic building blocks." Only when the dynamic, relativistic view is adopted does the paradox disappear. The particles are then seen as dynamic patterns, or processes, which involve a certain amount of energy appearing to us as their mass. In a collision process, the energy of the two colliding particles is redistributed to form a new pattern, and if it has been increased by a sufficient amount of kinetic energy, this new pattern may involve additional particles.

High-energy collisions of subatomic particles are the principal method used by physicists to study the properties of these particles, and particle physics is therefore also called "high-energy physics." The kinetic energies required for the collision experiments are achieved by means of huge particle accelerators,* enormous circular machines with circumferences of several miles in which protons are accelerated to velocities near the speed of light and are then made to collide with other protons or with neutrons. It is impressive that machines of that size are needed to study the world of the infinitely small. They are the supermicroscopes of our time.

Most of the particles created in these collisions live for only an extremely short time—much less than a millionth of a second—after which they disintegrate again into protons, neutrons, and electrons. In spite of their exceedingly short lifetime, these particles can not only be detected and their properties measured but are actually made to leave tracks which can be photographed! These particle tracks are produced in so-called bubble

* See photograph on page 1, showing an aerial view of the accelerator at Fermilab, near Batavia, Illinois, which has a circumference of four miles (photograph taken in 1971 while the laboratory was still under construction).

chambers in a manner similar to the way a jet plane makes a trail in the sky. The actual particles are many orders of magnitude smaller than the bubbles making up the tracks, but from the thickness and curvature of a track, physicists can identify the particle that caused it. The points from which several tracks emanate are points of particle collisions, and the curves are caused by magnetic fields which the experimenters use to identify the particles. The collisions of particles are our main experimental method to study their properties and interactions, and the beautiful lines, spirals, and curves traced by the particles in bubble chambers are thus of paramount importance for modern physics.

The high-energy scattering experiments of the past decades have shown us the dynamic and ever-changing nature of the particle world in the most striking way. Matter has appeared in these experiments as completely mutable. All particles can be transmuted into other particles; they can be created from energy and can vanish into energy. In this world, classical concepts like "elementary particle," "material substance," or "isolated object" have lost their meaning; the whole universe appears as a dynamic web of inseparable energy patterns. So far, we have not yet found a complete theory to describe this world of subatomic particles, but we do have several theoretical models which describe certain aspects of it very well. None of these models is free from mathematical difficulties, and they all contradict each other in certain ways, but all of them reflect the basic unity and the intrinsically dynamic character of matter. They show that the properties of a particle can be understood only in terms of its activity—of its interaction with the surrounding environment—and that the particle, therefore, cannot be seen as an isolated entity, but has to be understood as an integrated part of the whole.

Relativity theory has not only affected our conception of particles in a drastic way, but also our picture of the forces between these particles. In a relativistic description of particle interactions, the forces between the particles—that is, their mutual attraction or repulsion—are pictured as the exchange of other particles. This concept is very difficult to visualize. It is a consequence of the four-dimensional space-time character of the subatomic world and neither our intuition nor our language can deal with this image very well. Yet it is crucial for an understanding of subatomic phenomena. It links the forces between constituents of matter to the properties of other constituents of matter, and thus unifies the two concepts, force and matter, which had seemed to be so fundamentally different ever since the Greek atomists. Both force and matter are now seen to have their common origin in the dynamic patterns which we call particles.

The fact that particles interact through forces which manifest themselves as the exchange of other particles is yet another reason why the subatomic world cannot be decomposed into constituent parts. From the macroscopic level down to the nuclear level, the forces which hold things together are relatively weak, and it is a good approximation to say that things consist of constituent parts. Thus a grain of salt can be said to consist of salt molecules, the salt molecules of two kinds of atoms, those atoms to consist of nuclei and electrons, and the nuclei of protons and neutrons. At the particle level, however, it is no longer possible to see things that way.

In recent years, there has been an increasing amount of evidence that the protons and neutrons, too, are composite objects; but the forces holding them together are so strong or—what amounts to the same—the velocities acquired by the components are so high, that the relativistic picture has to be applied where the forces are also particles. Thus the distinction between the constituent particles and the particles making up the binding forces becomes blurred, and the approximation of an object consisting of constituent parts breaks down. The particle world cannot be decomposed into elementary components.

In modern physics, the universe is thus experienced as a dynamic, inseparable whole which always includes the observer in an essential way. In this experience, the traditional concepts of space and time, of isolated objects, and of cause and effect, lose their meaning. Such an experience, however, is very similar to that of the Eastern mystics. The similarity becomes apparent in quantum and relativity theory, and becomes even stronger in the "quantum-relativistic" models of subatomic physics where both these theories combine to produce the most striking parallels to Eastern mysticism.

Before spelling out these parallels in detail, I shall give a brief account of the schools of Eastern philosophy which are relevant to the comparison for the reader who is not familiar with them. They are the various schools in the religious philosophies of Hinduism, Buddhism, and Taoism. In the following five chapters, the historical background, characteristic features and philosophical concepts of these spiritual traditions will be described, the emphasis being on those aspects and concepts which will be important for the subsequent comparison with physics.

II
THE WAY OF EASTERN MYSTICISM

Shiva Mahesvara, Elephanta, India, eighth century A.D.

5
HINDUISM

For an understanding of any of the philosophies to be described, it is important to realize that they are religious in essence. Their main aim is the direct mystical experience of reality, and since this experience is religious by nature, they are inseparable from religion. More than for any other Eastern tradition, this is true for Hinduism, where the connection between philosophy and religion is particularly strong. It has been said that almost all thought in India is in a sense religious thought, and Hinduism has not only influenced, throughout many centuries, India's intellectual life, but almost completely determined her social and cultural life as well.

Hinduism cannot be called a philosophy, nor is it a well-defined religion. It is, rather, a large and complex socioreligious organism consisting of innumerable sects, cults, and philosophical systems and involving various rituals, ceremonies, and spiritual disciplines, as well as the worship of countless gods and goddesses. The many facets of this complex and yet persistent and powerful spiritual tradition mirror the geographical, racial, linguistic, and cultural complexities of India's vast subcontinent. The manifestations of Hinduism range from highly intellectual philosophies involving conceptions of fabulous range and depth, to the naïve and childlike ritual practices of the masses. If the majority of the Hindus are simple villagers who keep the popular religion alive in their daily worship, Hinduism has, on the other hand, brought forth a large number of outstanding spiritual teachers to transmit its profound insights.

The spiritual source of Hinduism lies in the *Vedas*, a collection of ancient scriptures written by anonymous sages, the so-called Vedic "seers." There are four *Vedas*, the oldest of them being the *Rig Veda*. Written in ancient Sanskrit, the sacred language of India, the *Vedas* have remained the highest religious authority for most sections of Hinduism. In India, any philosophical system that does not accept the authority of the *Vedas* is considered to be unorthodox.

Each of these *Vedas* consists of several parts which were composed at different periods, probably between 1500 and 500 B.C. The oldest parts are sacred hymns and prayers. Subsequent parts deal with sacrificial rituals connected with the Vedic hymns, and the last, called the *Upanishads*, elaborate their philosophical and practical content. The *Upanishads* contain the essence of Hinduism's spiritual message. They have guided and inspired India's greatest minds for the last twenty-five centuries, in accordance with the advice given in their verses:

> Taking as a bow the great weapon of the *Upanishad*,
> One should put upon it an arrow sharpened by meditation.
> Stretching it with a thought directed to the essence of That,
> Penetrate that Imperishable as the mark, my friend.[1]

The masses of the Indian people, however, have received the teachings of Hinduism not through the *Upanishads*, but through a large number of popular tales, collected in huge epics, which are the basis of the vast and colourful Indian mythology. One of those epics, the *Mahabharata*, contains India's favorite religious text, the beautiful spiritual poem of the *Bhagavad Gita*. The *Gita*, as it is commonly called, is a dialogue between the god Kṛishna and the warrior Arjuna, who is in great despair, being forced to combat his own kinsmen in the great family war which forms the main story of the *Mahabharata*. Krishna, disguised as Arjuna's charioteer, drives the chariot right between the two armies, and in this dramatic setting of the battlefield, he starts to reveal to Arjuna the most profound truths of Hinduism. As the god speaks, the realistic background of the war between the two families soon fades away, and it becomes clear that the battle of Arjuna is the spiritual battle of man, the battle of the warrior in search of enlightenment. Krishna himself advises Arjuna:

> Kill therefore with the sword of wisdom the doubt born of ignorance that lies in thy heart. Be one in self-harmony, in *Yoga*, and arise, great warrior, arise.[2]

The basis of Krishna's spiritual instruction, as of all Hinduism, is the idea that the multitude of things and events around us are but different manifestations of the same ultimate reality. This reality, called *Brahman*, is the unifying concept which gives Hinduism its essentially monistic character in spite of the worship of numerous gods and goddesses.

Brahman, the ultimate reality, is understood as the 'soul,' or inner essence, of all things. It is infinite and beyond all concepts; it cannot be comprehended by the intellect, nor can it be adequately described in words: "*Brahman*, beginningless, supreme: beyond what is and beyond what is not."[3]—"Incomprehensible is that supreme Soul, unlimited, unborn, not to be reasoned about, unthinkable."[4] Yet, people want to talk about this reality and the Hindu sages with their characteristic penchant for myth have pictured *Brahman* as divine and talk about it in mythological language. The various aspects of the Divine have been given the names of the various gods worshiped by the Hindus, but the scriptures make it clear that all these gods are but reflections of the one ultimate reality:

> This that people say, "Worship this god! Worship that god!"—one after another—this is his [*Brahman's*] creation indeed! And he himself is all the gods.[5]

The manifestation of *Brahman* in the human soul is called *Atman*, and the idea that *Atman* and *Brahman*, the individual and the ultimate reality, are one is the essence of the *Upanishads*:

> That which is the finest essence—this whole world has that as its soul. That is Reality. That is *Atman*. That art thou.[6]

The basic recurring theme in Hindu mythology is the

creation of the world by the self-sacrifice of God—'sacrifice' in the original sense of 'making sacred'—whereby God becomes the world which, in the end, becomes again God. This creative activity of the Divine is called *lila*, the play of God, and the world is seen as the stage of the divine play. Like most of Hindu mythology, the myth of *lila* has a strong magical flavor. *Brahman* is the great magician who transforms himself into the world and he performs this feat with his "magic creative power," which is the original meaning of *maya* in the *Rig Veda*. The word *maya*—one of the most important terms in Indian philosophy—has changed its meaning over the centuries. From the 'might,' or 'power,' of the divine actor and magician, it came to signify the psychological state of anybody under the spell of the magic play. As long as we confuse the myriad forms of the divine *lila* with reality, without perceiving the unity of *Brahman* underlying all these forms, we are under the spell of *maya*.

Maya, therefore, does not mean that the world is an illusion, as is often wrongly stated. The illusion merely lies in our point of view, if we think that the shapes and structures, things and events, around us are realities of nature, instead of realizing that they are concepts of our measuring and categorizing minds. *Maya* is the illusion of taking these concepts for reality, of confusing the map with the territory.

In the Hindu view of nature, then, all forms are relative, fluid and ever-changing *maya*, conjured up by the great magician of the divine play. The world of *maya* changes continuously, because the divine *lila* is a rhythmic, dynamic play. The dynamic force of the play is *karma*, another important concept of Indian thought. *Karma* means 'action.' It is the active principle of the play, the total universe in action, where everything is dynamically connected with everything else. In the words of the *Gita*, "*Karma* is the force of creation, wherefrom all things have their life."[7]

The meaning of *karma*, like that of *maya*, has been brought down from its original cosmic level to the human level, where it has acquired a psychological sense. As long as our view of the world is fragmented, as long as we are under the spell of *maya* and think that we are separated from our environment and can act independently, we are bound by *karma*. Being free from the bond of *karma* means to realize the unity and harmony of all nature, including man, and to act accordingly. The *Gita* is very clear on this point:

> All actions take place in time by the interweaving of the forces of nature, but the man lost in selfish delusion thinks that he himself is the actor.
>
> But the man who knows the relation between the forces of Nature and actions, sees how some forces of Nature work upon other forces of Nature, and becomes not their slave.[8]

To be free from the spell of *maya*, to break the bonds of *karma*, means to realize that all the phenomena we perceive with our senses are part of the same reality. It means to experience, concretely and personally, that everything, including our own self, is *Brahman*. This experience is called *moksha*, or 'liberation' in Hindu philosophy and it is the very essence of Hinduism.

Hinduism holds that there are innumerable ways of liberation. It would never expect all its followers to be able to approach the Divine in the same way, and therefore it provides different concepts, rituals, and spiritual exercises for different modes of awareness. The fact that many of these concepts or practices are contradictory does not worry the Hindus in the least, because they know that *Brahman* is beyond concepts and images anyway. From this attitude comes the great tolerance and inclusiveness which is characteristic of Hinduism.

The most intellectual school is the Vedanta which is based on the *Upanishads* and emphasizes *Brahman* as a nonpersonal, metaphysical concept, free from any mythological content. In spite of its high philosophical and intellectual level, however, the Vedantist way of liberation is very different from any school of Western philosophy, involving as it does daily meditation and other spiritual exercises to bring about the union with *Brahman*.

Another important and influential method of liberation is known as *yoga*, a word which means "to yoke," "to join," and which refers to the joining of the individual soul to *Brahman*. There are several schools, or 'paths' of *yoga* involving some basic physical training and various mental disciplines designed for people of different types and at different spiritual levels.

For the common Hindu, the most popular way of approaching the Divine is to worship it in the form of a personal god or goddess. The fertile Indian imagination has created literally thousands of deities which appear in innumerable manifestations. The three most worshiped divinities in India today are Shiva, Vishnu, and the Divine Mother. Shiva is one of the oldest Indian gods who can assume many forms. He is called *Mahesvara*, the Great Lord, when he is represented as the personification of the fullness of *Brahman* and he can also impersonate many single aspects of the Divine, his most celebrated appearance being the one as *Nataraja*, the King of Dancers. As the Cosmic Dancer, Shiva is the god of creation and destruction, who sustains through his dance the endless rhythm of the universe.

Vishnu, too, appears under many guises, one of them being the god Krishna of the *Bhagavad Gita*. In general, Vishnu's role is that of the preserver of the universe. The third divinity of this triad is Shakti, the Divine Mother, the archetypal goddess representing in her many forms the female energy of the universe.

Shakti also appears as Shiva's wife, and the two are often shown in passionate embraces in magnificent temple sculptures which radiate an extraordinary sensuousness of a degree completely unknown in any Western religious art. Contrary to most Western religions, sensuous pleasure has never been suppressed in Hinduism because the body has always been considered to be an integral part of the human being and not separated from the spirit. The Hindu, therefore, does not try to control the desires of the body by the conscious will, but aims at realizing himself with his whole being, body and mind. Hinduism has even developed a branch, the medieval Tantrism, where enlightenment is sought through a profound experience of sensual love 'in which each is both,' in accordance with the words of the *Upanishads*:

> As a man, when in the embrace of a beloved wife, knows nothing within or without, so this person,

when in the embrace of the intelligent Soul, knows nothing within or without.[9]

Shiva was closely associated with this medieval form of erotic mysticism, as were Shakti and numerous other female deities which exist in great numbers in Hindu mythology. This abundance of goddesses shows again that in Hinduism the physical and sensuous side of human nature, which has always been associated with the female, is a fully integrated part of the Divine. Hindu goddesses are not shown as holy virgins, but in sensual embraces of stunning beauty.

Stone sculpture, Khajuraho, India, circa A.D. 1000.

The Western mind is easily confused by the fabulous number of gods and goddesses which populate Hindu mythology in their various appearances and incarnations. To understand how the Hindus can cope with this multitude of divinities, we must be aware of the basic attitude of Hinduism that in substance all these divinities are identical. They are all manifestations of the same divine reality, reflecting different aspects of the infinite, omnipresent, and—ultimately—incomprehensible *Brahman*.

Face of the Buddha, India, fifth century A.D.

6
BUDDHISM

Buddhism has been, for many centuries, the dominant spiritual tradition in most parts of Asia, including the countries of Indochina, as well as Sri Lanka, Nepal, Tibet, China, Korea, and Japan. As with Hinduism in India, it has had a strong influence on the intellectual, cultural and artistic life of these countries. Unlike Hinduism, however, Buddhism goes back to a single founder, Siddhartha Gautama, the so-called "historic" Buddha. He lived in India in the middle of the sixth century B.C., during the extraordinary period that saw the birth of so many spiritual and philosophical geniuses: Confucius and Lao Tzu in China, Zarathustra in Persia, Pythagoras and Heraclitus in Greece.

If the flavor of Hinduism is mythological and ritualistic, that of Buddhism is definitely psychological. The Buddha was not interested in satisfying human curiosity about the origin of the world, the nature of the Divine, or similar questions. He was concerned exclusively with the human situation, with the suffering and frustrations of human beings. His doctrine, therefore, was not one of metaphysics, but one of psychotherapy. He pointed out the origin of human frustrations and the way to overcome them, taking up for this purpose the traditional Indian concepts of *maya, karma, nirvana,* etc., and giving them a fresh, dynamic, and directly relevant psychological interpretation.

After the Buddha's death, Buddhism developed into two main schools, the Hinayana and the Mahayana. The Hinayana, or Small Vehicle, is an orthodox school which sticks to the letter of the Buddha's teaching; whereas the Mahayana, or Great Vehicle, shows a more flexible attitude, believing that the spirit of the doctrine is more important than its original formulation. The Hinayana school established itself in Ceylon, Burma, and Thailand, whereas the Mahayana spread to Nepal, Tibet, China, and Japan and became, eventually, the more important of the two schools. In India itself, Buddhism was absorbed, after many centuries, by the flexible and assimilative Hinduism, and the Buddha was finally adopted as an incarnation of the many-faced god Vishnu.

As Mahayana Buddhism spread across Asia, it came into contact with peoples of many different cultures and mentalities who interpreted the Buddha's doctrine from their own point of view, elaborating many of its subtle points in great detail and adding their own original ideas. In this way they kept Buddhism alive over the centuries and developed highly sophisticated philosophies with profound psychological insights.

In spite of the high intellectual level of these philosophies, however, Mahayana Buddhism never loses itself in abstract speculative thought. As always in Eastern mysticism, the intellect is seen merely as a means to clear the way for the direct mystical experience, which Buddhists call the 'awakening.' The essence of this experience is to pass beyond the world of intellectual distinctions and opposites to reach the world of *acintya*, the unthinkable, where reality appears as undivided and undifferentiated 'suchness.'

This was the experience Siddhartha Gautama had one night, after seven years of strenuous discipline in the forests. Sitting in deep meditation under the celebrated Bodhi Tree, the Tree of Enlightenment, he suddenly obtained the final and definite clarification of all his searches and doubts in the act of 'unexcelled, complete awakening' which made him the *Buddha*, that is, 'the Awakened.' For the Eastern world, the Buddha's image in the state of meditation is as significant as the image of the crucified Christ for the West, and has inspired countless artists all over Asia who have created magnificent sculptures of meditating Buddhas.

According to Buddhist tradition, the Buddha went to the Deer Park of Benares immediately after his awakening to preach his doctrine to his former fellow hermits. He expressed it in the celebrated form of the Four Noble

Truths, a compact presentation of the essential doctrine which is not unlike the statement of a physician, who first identifies the cause of humanity's sickness, then affirms that the sickness can be cured, and finally prescribes the remedy.

The First Noble Truth states the outstanding characteristic of the human situation, *duhkha*, which is suffering or frustration. This frustration comes from our difficulty in facing the basic fact of life, that everything around us is impermanent and transitory. "All things arise and pass away,"[1] said the Buddha, and the notion that flow and change are basic features of nature lies at the root of Buddhism. Suffering arises, in the Buddhist view, whenever we resist the flow of life and try to cling to fixed forms which are all *maya*, whether they are things, events, people, or ideas. This doctrine of impermanence includes also the notion that there is no ego, no self which is the persistent subject of our varying experiences. Buddhism holds that the idea of a separate individual self is an illusion, just another form of *maya*, an intellectual concept which has no reality. To cling to this concept leads to the same frustration as adherence to any other fixed category of thought.

The Second Noble Truth deals with the cause of all suffering, *trishna*, which is clinging, or grasping. It is the futile grasping of life based on a wrong point of view which is called *avidya*, or ignorance, in Buddhist philosophy. Out of this ignorance, we divide the perceived world into individual and separate things and thus attempt to confine the fluid forms of reality in fixed categories created by the mind. As long as this view prevails, we are bound to experience frustration after frustration. Trying to cling to things which we see as firm and persistent, but which in fact are transient and ever-changing, we are trapped in a vicious circle where every action generates further action and the answer to each question poses new questions. This vicious circle is known in Buddhism as *samsara*, the round of birth-and-death, and it is driven by *karma*, the never-ending chain of cause and effect.

The Third Noble Truth states that the suffering and frustration can be ended. It is possible to transcend the vicious circle of *samsara*, to free oneself from the bondage of *karma*, and to reach a state of total liberation called *nirvana*. In this state, the false notions of a separate self have forever disappeared, and the oneness of all life has become a constant sensation. *Nirvana* is the equivalent of *moksha* in Hindu philosophy and, being a state of consciousness beyond all intellectual concepts, it defies further description. To reach *nirvana* is to attain awakening, or Buddhahood.

The Fourth Noble Truth is the Buddha's prescription to end all suffering, the Eightfold Path of self-development which leads to the state of Buddhahood. The first two sections of this path, as already mentioned, are concerned with right seeing and right knowing; that is, with the clear insight into the human situation that is the necessary starting point. The next four sections deal with right action. They give the rules for the Buddhist way of life, which is a Middle Way between opposite extremes. The last two sections are concerned with right awareness and right meditation and describe the direct mystical experience of reality that is the final goal.

The Buddha did not develop his doctrine into a consistent philosophical system, but regarded it as a means to achieve enlightenment. His statements about the world were confined to emphasizing the impermanence of all 'things.' He insisted on freedom from spiritual authority, including his own, saying that he could only show the way to Buddhahood, and that it was up to every individual to tread this way to the end through his or her own efforts. The Buddha's last words on his deathbed are characteristic of his world view and of his attitude as a teacher. "Decay is inherent in all compounded things," he said before passing away: "Strive on with diligence."[2]

In the first few centuries after the Buddha's death, several Great Councils were held by the leading monks of the Buddhist order at which the entire teaching was recited aloud and differences in interpretation were settled. At the fourth of these councils, which took place on the island of Ceylon (Sri Lanka) in the first century A.D., the memorized doctrine, which had been passed on orally for more than five hundred years, was for the first time recorded in writing. This record, written in the Pali language, is known as the Pali Canon and forms the basis of the orthodox Hinayana school. The Mahayana school, on the other hand, is based on a number of so-called *sutras*, scriptures of huge dimensions, which were written in Sanskrit one or two hundred years later and present the Buddha's teaching in a much more elaborate and subtle way than the Pali Canon.

The Mahayana school calls itself the Great Vehicle of Buddhism because it offers its adherents a great variety of methods, or 'skillful means' to attain Buddhahood. These range from doctrines emphasizing religious faith in the teachings of the Buddha, to elaborate philosophies involving concepts which come very close to modern scientific thought.

The first expounder of the Mahayana doctrine, and one of the deepest thinkers among the Buddhist patriarchs, was Ashvaghosha, who lived in the first century A.D. He spelled out the fundamental thoughts of Mahayana Buddhism—in particular those relating to the Buddhist concept of 'suchness'—in a small book called *The Awakening of Faith*. This lucid and extremely beautiful text, which reminds one of the *Bhagavad Gita* in many ways, constitutes the first representative treatise on the Mahayana doctrine and has become a principal authority for all schools of Mahayana Buddhism.

Ashvaghosha probably had a strong influence on Nagarjuna, the most intellectual Mahayana philosopher, who used a highly sophisticated dialectic to show the limitations of all concepts of reality. With brilliant arguments he demolished the metaphysical propositions of his time and thus demonstrated that reality, ultimately, cannot be grasped with concepts and ideas. Hence, he gave it the name *sunyata*, 'the void,' or 'emptiness,' a term which is equivalent to Ashvaghosha's *tathata*, or 'suchness'; when the futility of all conceptual thinking is recognized, reality is experienced as pure suchness.

Nagarjuna's statement that the essential nature of reality is emptiness is thus far from being the nihilist statement for which it is often taken. It merely means that all concepts about reality formed by the human mind are ultimately void. Reality, or Emptiness, itself is not a state of mere nothingness, but is the very source of all life and the essence of all forms.

The views of Mahayana Buddhism presented so far reflect its intellectual, speculative side. This, however, is only one side of Buddhism. Complementary to it is the Buddhist's religious consciousness which involves faith, love and compassion. True enlightened wisdom (*bodhi*) is seen in the Mahayana as being composed of two elements which D. T. Suzuki has called the "two pillars supporting the great edifice of Buddhism." They are *prajna*, which is transcendental wisdom, or intuitive intelligence, and *Karuna*, which is love or compassion.

Accordingly, the essential nature of all things is described in Mahayana Buddhism not only by the abstract metaphysical terms Suchness and Void, but also by the term *Dharmakaya*, the 'Body of Being,' which describes reality as it appears to the Buddhist's religious consciousness. The *Dharmakaya* is similar to the *Brahman* in Hinduism. It pervades all material things in the universe and is also reflected in the human mind as *bodhi*, the enlightened wisdom. It is thus spiritual and material at the same time.

The emphasis on love and compassion as essential parts of wisdom has found its strongest expression in the ideal of the Bodhisattva, one of the characteristic developments of Mahayana Buddhism. A Bodhisattva is a highly evolved human being on the way to becoming a Buddha, who is not seeking enlightenment for himself alone, but has vowed to help all other beings achieve Buddhahood before he enters into *nirvana*. The origin of this idea lies in the decision of the Buddha—presented in Buddhist tradition as a conscious and not at all easy decision—not simply to enter *nirvana*, but to return to the world in order to show the path to salvation to his fellow human beings. The Bodhisattva ideal is also consistent with the Buddhist doctrine of nonego, because if there is no separate individual self, the idea of one individual entering *nirvana* alone obviously does not make much sense.

The element of faith, finally, is emphasized in the so-called Pure Land school of Mahayana Buddhism. The basis of this school is the Buddhist doctrine that the original nature of all human beings is that of a Buddha, and it holds that in order to enter *nirvana*, or the 'Pure

Land,' all one has to do is to have faith in one's original Buddha nature.

The culmination of Buddhist thought has been reached, according to many authors, in the so-called *Avatamsaka* school which is based on the *sutra* of the same name. This *sutra* is regarded as the core of Mahayana Buddhism and is praised by Suzuki in the most enthusiastic words:

> As to the *Avatamsaka-sutra*, it is really the consummation of Buddhist thought, Buddhist sentiment, and Buddhist experience. To my mind, no religious literature in the world can ever approach the grandeur of conception, the depth of feeling, and the gigantic scale of composition as attained in this *sutra*. It is the eternal fountain of life from which no religious mind will turn back athirst or only partially satisfied.[3]

It was this *sutra* which stimulated Chinese and Japanese minds more than anything else, when Mahayana Buddhism spread across Asia. The contrast between the Chinese and Japanese, on the one hand, and the Indians, on the other, is so great that they have been said to represent two poles of the human mind. Whereas the former are practical, pragmatic, and socially minded, the latter are imaginative, metaphysical, and transcendental. When the Chinese and Japanese philosophers began to translate and interpret the *Avatamsaka*, one of the greatest scriptures produced by the Indian religious genius, the two poles combined to form a new dynamic unity and the outcome were the *Hua-yen* philosophy in China and the *Kegon* philosophy in Japan which constitute, according to Suzuki, "the climax of Buddhist thought which has been developing in the Far East for the last two thousand years."[4]

The central theme of the *Avatamsaka* is the unity and interrelation of all things and events; a conception which is not only the very essence of the Eastern world-view, but also one of the basic elements of the world-view emerging from modern physics. It will therefore be seen that the *Avatamsaka Sutra*, this ancient religious text, offers the most striking parallels to the models and theories of modern physics.

Inkcake by Ch'eng Chung-fang, seventeenth century.

7
CHINESE THOUGHT

When Buddhism arrived in China, around the first century A.D., it encountered a culture which was more than two thousand years old. In this ancient culture, philosophical thought had reached its culmination during the late Chou period (c. 500–221 B.C.), the golden age of Chinese philosophy, and from then on had always been held in the highest esteem.

From the beginning, this philosophy had two complementary aspects. The Chinese being practical people with a highly developed social consciousness, all their philosophical schools were concerned, in one way or the other, with life in society, with human relations, moral values and government. This, however, is only one aspect of Chinese thought. Complementary to it is that corresponding to the mystical side of the Chinese character, which demanded that the highest aim of philosophy should be to transcend the world of society and everyday life and to reach a higher plane of consciousness. This is the plane of the sage, the Chinese ideal of the enlightened man who has achieved mystical union with the universe.

The Chinese sage, however, does not dwell exclusively on this high spiritual plane, but is equally concerned with worldly affairs. He unifies in himself the two complementary sides of human nature—intuitive wisdom and practical knowledge, contemplation and social action—which the Chinese have associated with the images of the sage and of the king. Fully realized human beings, in the words of Chuang Tzu, "by their stillness become sages, by their movement kings."[1]

During the sixth century B.C., the two sides of Chinese philosophy developed into two distinct philosophical schools, Confucianism and Taoism. Confucianism was the philosophy of social organization, of common sense and practical knowledge. It provided Chinese society with a system of education and with strict conventions of social etiquette. One of its main purposes was to form an ethical basis for the traditional Chinese family system with its complex structure and its rituals of ancestor worship. Taoism, on the other hand, was concerned primarily with the observation of nature and the discovery of its Way, or *Tao*. Human happiness, according to the Taoists, is achieved when men follow the natural order, acting spontaneously and trusting their intuitive knowledge.

These two trends of thought represent opposite poles in Chinese philosophy, but in China they were always seen as poles of one and the same human nature, and thus as complementary. Confucianism was generally emphasized in the education of children who had to learn the rules and conventions necessary for life in society, whereas Taoism used to be pursued by older people in order to regain and develop the original spontaneity which had been destroyed by social conventions. In the eleventh and twelfth centuries, the Neo-Confucian school attempted a synthesis of Confucianism, Buddhism and Taoism, which culminated in the philosophy of Chu Hsi, one of the greatest of all Chinese thinkers. Chu Hsi was an outstanding philosopher who combined Confucian scholarship with a deep understanding of Buddhism and Taoism, and incorporated elements of all three traditions in his philosophical synthesis.

Confucianism derives its name from Kung Fu Tzu, or Confucius, a highly influential teacher with a large number of students who saw his main function as transmitting the ancient cultural heritage to his disciples. In doing so, however, he went beyond a simple transmission of knowledge for he interpreted the traditional ideas according to his own moral concepts. His teachings were based on the so-called Six Classics, ancient books of philosophical thought, rituals, poetry, music, and history, which represented the spiritual and cultural heritage of the "holy sages" of China's past. Chinese tradition has associated Confucius with all of these works, either as author, commentator, or editor; but according to modern scholarship he was neither the author, commentator, nor even the editor of any of the Classics. His own ideas became known through the *Lun Yü*, or Confucian Analects, a collection of aphorisms which was compiled by some of his disciples.

The originator of Taoism was Lao Tzu, whose name literally means "The Old Master" and who was, according to tradition, an older contemporary of Confucius. He is said to have been the author of a short book of aphorisms which is considered as the main Taoist scripture. In China, it is generally just called the *Lao-tzu*, and in the West it is usually known as the *Tao Te Ching*, the *Classic of the Way and Power*, a name which was given to it in later times. I have already mentioned the paradoxical style and the powerful and poetic language of this book which Joseph Needham considers to be "without exception the most profound and beautiful work in the Chinese language."[2]

The second important Taoist book is the *Chuang-tzu*, a much larger book than the *Tao Te Ching*, whose author, Chuang Tzu, is said to have lived about two hundred years after Lao Tzu. According to modern scholarship, however, the *Chuang-tzu*, and probably also the *Lao-tzu*, cannot be seen as the work of a single author, but rather constitute a collection of Taoist writings compiled by different authors at different times.

Both the Confucian Analects and the *Tao Te Ching* are written in the compact suggestive style which is typical of the Chinese way of thinking. The Chinese mind was not given to abstract logical thinking and developed a language which is very different from that which evolved in the West. Many of its words could be used as nouns, adjectives, or verbs, and their sequence was determined not so much by grammatical rules as by the emotional content of the sentence. The classical Chinese word was very different from an abstract sign representing a clearly delineated concept. It was rather a sound symbol which had strong suggestive powers, bringing to mind an indeterminate complex of pictorial images and emotions. The intention of the speaker was not so much to express an intellectual idea, but rather to affect and influence the listener. Correspondingly, the written character was not just an abstract sign, but was an organic pattern—a "gestalt"—which preserved the full complex of images and the suggestive power of the word.

Since the Chinese philosophers expressed themselves in a language which was so well suited for their way of thinking, their writings and sayings could be short and inarticulate, and yet rich in suggestive images. It is clear that much of this imagery must be lost in an English translation. A translation of a sentence from the *Tao Te Ching*, for example, can only render a small part of the rich complex of ideas contained in the original, which is why different translations from this controversial book often look like totally different texts. As Fung Yu-Lan has said, "It needs a combination of all the translations already made and many others not yet made, to reveal the richness of the *Lao-tzu* and the Confucian Analects in their original form."[3]

The Chinese, like the Indians, believed that there is an ultimate reality which underlies and unifies the multiple things and events we observe:

> There are the three terms—"complete," "all-embracing," "the whole." These names are different, but the reality sought in them is the same: referring to the One thing.[4]

They called this reality the *Tao*, which originally meant 'the Way.' It is the way, or process, of the universe, the order of nature. In later times, the Confucianists gave it a different interpretation. They talked about the *Tao* of man, or the *Tao* of human society, and understood it as the right way of life in a moral sense.

In its original cosmic sense, the *Tao* is the ultimate, undefinable reality and as such it is the equivalent of the Hinduist *Brahman* and the Buddhist *Dharmakaya*. It differs from these Indian concepts, however, by its intrinsically dynamic quality which, in the Chinese view, is the essence of the universe. The *Tao* is the cosmic process in which all things are involved; the world is seen as a continuous flow and change.

Indian Buddhism, with its doctrine of impermanence, had quite a similar view, but it took this view merely as the basic premise of the human situation and went on to elaborate its psychological consequences. The Chinese, on the other hand, not only believed that flow and change were the essential features of nature, but also that there are constant patterns in these changes, to be observed by man. The sage recognizes these patterns and directs his actions according to them. In this way, he becomes 'one with the *Tao*,' living in harmony with nature and succeeding in everything he undertakes. In the words of Huai Nan Tzu, a philosopher of the second century B.C.:

> He who conforms to the course of the *Tao*, following the natural processes of Heaven and Earth, finds it easy to manage the whole world.[5]

What, then, are the patterns of the cosmic Way which man has to recognize? The principal characteristic of the *Tao* is the cyclic nature of its ceaseless motion and change. "Returning is the motion of the *Tao*," says Lao Tzu, and "Going far means returning."[6] The idea is that all developments in nature, those in the physical world as well as those of human situations, show cyclic patterns of coming and going, of expansion and contraction.

This idea was no doubt deduced from the movements of the sun and moon and from the change of the seasons, but it was then also taken as a rule of life. The Chinese believe that whenever a situation develops to its extreme, it is bound to turn around and become its opposite. This basic belief has given them courage and perseverance in times of distress and has made them cautious and modest in times of success. It has led to the doctrine of the golden mean in which both Taoists and Confucianists believe. "The sage," says Lao Tzu, "avoids excess, extravagance, and indulgence."[7]

In the Chinese view, it is better to have too little than to have too much, and better to leave things undone than to overdo them, because although one may not get very far this way, one is certain to go in the right direction. Just as the man who wants to go farther and farther East will end up in the West, those who accumulate more and more money in order to increase their wealth will end up being poor. Modern industrial society which is continuously trying to increase the "standard of living" and thereby decreases the quality of life for all its members is an eloquent illustration of this ancient Chinese wisdom.

The idea of cyclic patterns in the motion of the *Tao* was given a definite structure by the introduction of the polar opposites *yin* and *yang*. They are the two poles which set the limits for the cycles of change:

> The *yang* having reached its climax retreats in favor of the *yin*; the *yin* having reached its climax retreats in favor of the *yang*.[8]

In the Chinese view, all manifestations of the *Tao* are generated by the dynamic interplay of these two polar forces. This idea is very old and many generations worked on the symbolism of the archetypal pair *yin* and *yang* until it became the fundamental concept of Chinese thought. The original meaning of the words *yin* and *yang* was that of the shady and sunny sides of a mountain, a meaning which gives a good idea of the relativity of the two concepts:

> That which lets now the dark, now the light appear is *Tao*.[9]

From the very early times, the two archetypal poles of nature were represented not only by bright and dark, but also by male and female, firm and yielding, above and below. *Yang*, the strong, male, creative power, was associated with Heaven, whereas *yin*, the dark, receptive, female and maternal element, was represented by the Earth. Heaven is above and full of movement, the Earth—in the old geocentric view—is below and resting, and thus *yang* came to symbolize movement and *yin* rest. In the realm of thought, *yin* is the complex, female, intuitive mind, *yang* the clear and rational male intellect. *Yin* is the quiet, contemplative stillness of the sage, *yang* the strong, creative action of the king.

The dynamic character of *yin* and *yang* is illustrated by the ancient Chinese symbol called *T'ai-chi T'u*, or 'Diagram of the Supreme Ultimate':

This diagram is a symmetric arrangement of the dark *yin* and the bright *yang*, but the symmetry is not static. It is a rotational symmetry suggesting, very forcefully, a continuous cyclic movement:

> The *yang* returns cyclically to its beginning; the *yin* attains its maximum and gives place to the *yang*.[10]

The two dots in the diagram symbolize the idea that each time one of the two forces reaches its extreme, it contains in itself already the seed of its opposite.

The pair of *yin* and *yang* is the grand leitmotiv that permeates Chinese culture and determines all features of the traditional Chinese way of life. "Life," says Chuang Tzu, "is the blended harmony of the *yin* and *yang*."[11] As a nation of farmers, the Chinese had always been familiar with the movements of the sun and moon and with the change of the seasons. Seasonal changes and the resulting phenomena of growth and decay in organic nature were thus seen by them as the clearest expressions of the interplay between *yin* and *yang*, between the cold and dark winter and the bright and hot summer. The seasonal interplay of the two opposites is also reflected in the food we eat which contains elements of *yin* and *yang*. A healthy diet consists, for the Chinese, in balancing these *yin* and *yang* elements.

Traditional Chinese medicine, too, is based on the balance of *yin* and *yang* in the human body, and any illness is seen as a disruption of this balance. The body is divided into *yin* and *yang* parts. Globally speaking, the inside of the body is *yang*, the body surface is *yin*; the back is *yang*, the front is *yin*; inside the body, there are *yin* and *yang* organs. The balance between all these parts is maintained by a continuous flow of *ch'i*, or vital energy, along a system of "meridians" which contain the acupuncture points. Each organ has a meridian associated with it in such a way that *yang* meridians belong to *yin* organs and vice versa. Whenever the flow between the *yin* and *yang* is blocked, the body falls ill, and the illness is cured by sticking needles into the acupuncture points to stimulate and restore the flow of *ch'i*.

The interplay of *yin* and *yang*, the primordial pair of opposites, appears thus as the principle that guides all the movements of the Tao, but the Chinese did not stop there. They went on to study various combinations of *yin* and *yang* which they developed into a system of cosmic archetypes. This system is elaborated in the *I Ching*, or *Book of Changes*.

The *Book of Changes* is the first among the six Confucian Classics and must be considered as a work which lies at the very heart of Chinese thought and culture. The authority and esteem it has enjoyed in China throughout thousands of years is comparable only to those of sacred scriptures, like the *Vedas* or the Bible, in other cultures. The noted sinologue Richard Wilhelm begins the introduction to his translation of the book with the following words:

> The Book of Changes—*I Ching* in Chinese—is unquestionably one of the most important books in the world's literature. Its origin goes back to mythical antiquity, and it has occupied the attention of the most eminent scholars of China down to the present day. Nearly all that is greatest and most significant in the three thousand years of Chinese cultural history has either taken its inspiration from this book, or has exerted an influence on the interpretation of its text. Therefore it may safely be said that the seasoned wisdom of thousands of years has gone into the making of the *I Ching*.[12]

The *Book of Changes* is thus a work that has grown organically over thousands of years and consists of many layers stemming from the most important periods of Chinese thought. The starting point of the book was a collection of sixty-four figures, or "hexagrams," of the following type, which are based on the *yin-yang* symbolism and were used as oracles. Each hexagram consists of six lines which may be either broken (*yin*) or unbroken (*yang*), the sixty-four of them constituting all possible combinations of that kind. These hexagrams, which will be

discussed in greater detail later on, were considered as cosmic archetypes representing the patterns of the *Tao* in nature and in human situations. Each of them was given a title and was supplemented with a brief text, called the Judgment, to indicate the course of action appropriate to the cosmic pattern in question. The so-called Image is another brief text, added at a later date, which elaborates the meaning of the hexagram in a few, often extremely poetic lines. A third text interprets each of the hexagram's six lines in a language charged with mythical images which are often difficult to understand.

These three categories of texts form the basic parts of the book which were used for divination. An elaborate ritual involving fifty yarrow stalks was used to determine the hexagram corresponding to the personal situation of the questioner. The idea was to make the cosmic pattern of that moment visible in the hexagram and to learn from the oracle which course of action was appropriate to it:

> In the Changes there are images to reveal, there are judgments appended in order to interpret, good fortune and misfortune are determined in order to decide.[13]

The purpose of consulting the *I Ching* was thus not merely to know the future, but rather to discover the disposition of the present situation so that proper action could be taken. This attitude lifted the *I Ching* above the level of an ordinary book of soothsaying and made it a book of wisdom.

The use of the *I Ching* as a book of wisdom is, in fact, of far greater importance than its use as an oracle. It has inspired the leading minds of China throughout the ages, among them Lao Tzu, who drew some of his profoundest aphorisms from this source. Confucius studied it intensively and most of the commentaries on the text which make up the later strata of the book go back to his school. These commentaries, the so-called Ten Wings, combine the structural interpretation of the hexagrams with philosophical explanations.

At the center of the Confucian commentaries, as of the entire *I Ching*, is the emphasis on the dynamic aspect of all phenomena. The ceaseless transformation of all things and situations is the essential message of the *Book of Changes*:

> The Changes is a book
> From which one may not hold aloof.
> Its *tao* is forever changing—
> Alteration, movement without rest,
> Flowing through the six empty places,
> Rising and sinking without fixed law,
> Firm and yielding transform each other.
> They cannot be confined within a rule,
> It is only change that is at work here.[14]

8
TAOISM

Of the two main Chinese trends of thought, Confucianism and Taoism, the latter is the one which is mystically oriented and thus more relevant for our comparison with modern physics. Like Hinduism and Buddhism, Taoism is interested in intuitive wisdom, rather than in rational knowledge. Acknowledging the limitations and the relativity of the world of rational thinking, Taoism is, basically, a way of liberation from this world and is, in this respect, comparable to the ways of Yoga or Vedanta in Hinduism, or to the Eightfold Path of the Buddha. In the context of Chinese culture, the Taoist liberation meant, more specifically, a liberation from the strict rules of convention.

Mistrust of conventional knowledge and reasoning is stronger in Taoism than in any other school of Eastern philosophy. It is based on the firm belief that the human intellect can never comprehend the *Tao*. In the words of Chuang Tzu,

> The most extensive knowledge does not necessarily know it; reasoning will not make men wise in it. The sages have decided against both these methods.[1]

Chuang Tzu's book is full of passages reflecting the Taoist's contempt of reasoning and argumentation. Thus he says,

> A dog is not reckoned good because he barks well, and a man is not reckoned wise because he speaks skillfully.[2]

and

> Disputation is a proof of not seeing clearly.[3]

Logical reasoning was considered by the Taoists as part of the artificial world of man, together with social etiquette and moral standards. They were not interested in this world at all, but concentrated their attention fully on the observation of nature in order to discern the 'characteristics of the *Tao*.' Thus they developed an attitude which was essentially scientific and only their deep mistrust in the analytic method prevented them from constructing proper scientific theories. Nevertheless, the careful observation of nature, combined with a strong mystical intuition, led the Taoist sages to profound insights which are confirmed by modern scientific theories.

One of the most important insights of the Taoists was the realization that transformation and change are essential features of nature. A passage in the *Chuang-tzu* shows clearly how the fundamental importance of change was discerned by observing the organic world:

> In the transformation and growth of all things, every bud and feature has its proper form. In this we have their gradual maturing and decay, the constant flow of transformation and change.[4]

The Taoists saw all changes in nature as manifestations of the dynamic interplay between the polar opposites *yin* and *yang*, and thus they came to believe that any pair of opposites constitutes a polar relationship where each of the two poles is dynamically linked to the other. For the Western mind, this idea of the implicit unity of all opposites is extremely difficult to accept. It seems most paradoxical to us that experiences and values which we had always believed to be contrary should be, after all, aspects of the same thing. In the East, however, it has always been considered as essential for attaining enlightenment to go 'beyond earthly opposites,'[5] and in China the polar relationship of all opposites lies at the very basis of Taoist thought. Thus Chuang Tzu says,

> The "this" is also "that." The "that" is also "this." . . . That the "that" and the "this" cease to be opposites is the very essence of *Tao*. Only this essence, an axis as it were, is the center of the circle responding to the endless changes.[6]

From the notion that the movements of the *Tao* are a continuous interplay between opposites, the Taoists deduced two basic rules for human conduct. Whenever you want to achieve anything, they said, you should start with its opposite. Thus Lao Tzu:

> In order to contract a thing, one should surely expand it first.
> In order to weaken, one will surely strengthen first.
> In order to overthrow, one will surely exalt first.
> "In order to take, one will surely give first."
> This is called subtle wisdom.[7]

On the other hand, whenever you want to retain anything, you should admit in it something of its opposite:

> Be bent, and you will remain straight.
> Be vacant, and you will remain full.
> Be worn, and you will remain new.[8]

This is the way of life of the sage who has reached a higher point of view, a perspective from which the relativity and polar relationship of all opposites are clearly perceived. These opposites include, first and foremost, the concepts of good and bad which are interrelated in the same way as *yin* and *yang*. Recognizing the relativity of good and bad, and thus of all moral standards, the Taoist sage does not strive for the good but rather tries to maintain a dynamic balance between good and bad. Chuang Tzu is very clear on this point:

> The sayings, "Shall we not follow and honor the right and have nothing to do with the wrong?" and "Shall we not follow and honor those who secure good government and have nothing to do with those who produce disorder?" show a want of acquaintance with the principles of Heaven and Earth and with the different qualities of things. It is like following and honoring Heaven and taking no account of Earth; it is like following and honoring the *yin* and taking no account of the *yang*. It is clear that such a course cannot be pursued.[9]

It is amazing that, at the same time when Lao Tzu and his followers developed their world view, the essential features of this Taoist view were taught also in Greece, by a man whose teachings are known to us only in fragments and who was, and still is, very often misunderstood. This Greek "Taoist" was Heraclitus of Ephesus. He shared with Lao Tzu not only the emphasis on continuous change, which he expressed in his famous saying "Everything flows," but also the notion that all changes are cyclic. He compared the world order to "an ever-living fire, kindling in measures and going out in measures,"[10] an image which is indeed very similar to the Chinese idea of the *Tao* manifesting itself in the cyclic interplay of *yin* and *yang*.

It is easy to see how the concept of change as a dynamic interplay of opposites led Heraclitus, like Lao Tzu, to the discovery that all opposites are polar and thus united. "The way up and down is one and the same," said the Greek, and "God is day night, winter summer, war peace, satiety hunger."[11] Like the Taoists, he saw any pair of opposites as a unity and was well aware of the relativity of all such concepts. Again the words of Heraclitus—"Cold things warm themselves, warm cools, moist dries, parched is made wet"[12]—remind us strongly of those of Lao Tzu, "Easy gives rise to difficult . . . resonance harmonizes sound, after follows before."[13]

It is surprising that the great similarity between the world views of those two sages of the sixth century B.C. is not generally known. Heraclitus is often mentioned in connection with modern physics, but hardly ever in connection with Taoism. And yet it is this connection which shows best that his world view was that of a mystic and thus, in my opinion, puts the parallels between his ideas and those of modern physics in the right perspective.

When we talk about the Taoist concept of change, it is important to realize that this change is not seen as occurring as a consequence of some force, but rather as a tendency which is innate in all things and situations. The movements of the *Tao* are not forced upon it, but occur naturally and spontaneously. Spontaneity is the *Tao*'s principle of action, and since human conduct should be modeled on the operation of the *Tao*, spontaneity should also be characteristic of all human actions. Acting in harmony with nature thus means for the Taoists acting spontaneously and according to one's true nature. It means trusting one's intuitive intelligence, which is innate in the human mind just as the laws of change are innate in all things around us.

The actions of the Taoist sage thus arise out of his intuitive wisdom, spontaneously and in harmony with his environment. He does not need to force himself, or anything around him, but merely adapts his actions to the movements of the *Tao*. In the words of Huai Nan Tzu,

> Those who follow the natural order flow in the current of the *Tao*.[14]

Such a way of acting is called *wu-wei* in Taoist philosophy; a term which means literally "nonaction," and which Joseph Needham translates as "refraining from activity contrary to nature," justifying this interpretation with a quotation from the *Chuang-tzu*:

> Nonaction does not mean doing nothing and keeping silent. Let everything be allowed to do what it naturally does, so that its nature will be satisfied.[15]

If one refrains from acting contrary to nature or, as Needham says, from "going against the grain of things," one is in harmony with the *Tao* and thus one's actions will be successful. This is the meaning of Lao Tzu's seemingly so puzzling words, "By nonaction everything can be done."[16]

The contrast of *yin* and *yang* is not only the basic ordering principle throughout Chinese culture, but is also reflected in the two dominant trends of Chinese thought. Confucianism was rational, masculine, active and dominating. Taoism, on the other hand, emphasized all that was intuitive, feminine, mystical, and yielding. "Not knowing that one knows is best," says Lao Tzu, and "The sage carries on his business without action and gives his teachings without words."[17] The Taoists believed that by displaying the feminine, yielding qualities of human nature, it was easiest to lead a perfectly balanced life in harmony with the Tao. Their ideal is best summed up in a passage from the *Chuang-tzu* which describes a kind of Taoist paradise:

The men of old, while the chaotic condition was yet undeveloped, shared the placid tranquillity which belonged to the whole world. At that time the *yin* and *yang* were harmonious and still; their resting and movement proceeded without any disturbance; the four seasons had their definite times; not a single thing received any injury, and no living being came to a premature end. Men might be possessed of the faculty of knowledge, but they had no occasion for its use. This was what is called the state of perfect unity. At this time, there was no action on the part of anyone—but a constant manifestation of spontaneity.[18]

9
ZEN

When the Chinese mind came in contact with Indian thought in the form of Buddhism, around the first century A.D., two parallel developments took place. On the one hand, the translation of the Buddhist *sutras* stimulated Chinese thinkers and led them to interpret the teachings of the Indian Buddha in the light of their own philosophies. Thus arose an immensely fruitful exchange of ideas which culminated, as already mentioned, in the *Hua-yen*

Calligraphy by Ryokwan, eighteenth century.

(Sanskrit: *Avatamsaka*) school of Buddhism in China and in the *Kegon* school in Japan.

On the other hand, the pragmatic side of the Chinese mentality responded to the impact of Indian Buddhism by concentrating on its practical aspects and developing them into a special kind of spiritual discipline which was given the name *Ch'an*, a word usually translated as "meditation." This *Ch'an* philosophy was eventually adopted by Japan, around A.D. 1200, and has been cultivated there, under the name of Zen, as a living tradition up to the present day.

Zen is thus a unique blend of the philosophies and idiosyncrasies of three different cultures. It is a way of life which is typically Japanese, and yet it reflects the mysticism of India, the Taoists' love of naturalness and spontaneity and the thorough pragmatism of the Confucian mind.

In spite of its rather special character, Zen is purely Buddhistic in its essence because its aim is no other than that of the Buddha himself: the attainment of enlightenment, an experience known in Zen as *satori*. The enlightenment experience is the essence of all schools of Eastern philosophy, but Zen is unique in that it concentrates exclusively on this experience and is not interested in any further interpretations. In the words of Suzuki, "Zen is discipline in enlightenment." From the standpoint of Zen, the awakening of the Buddha and the Buddha's teaching that everybody has the potential of attaining this awakening are the essence of Buddhism. The rest of the doctrine, as expounded in the voluminous *sutras*, is seen as supplementary.

The experience of Zen is thus the experience of *satori*, and since this experience, ultimately, transcends all categories of thought, Zen is not interested in any abstraction or conceptualization. It has no special doctrine or philosophy, no formal creeds or dogmas, and it asserts that this freedom from all fixed beliefs makes it truly spiritual.

More than any other school of Eastern mysticism, Zen is convinced that words can never express the ultimate truth. It must have inherited this conviction from Taoism, which showed the same uncompromising attitude. "If one asks about the *Tao* and another answers him," said Chuang Tzu, "neither of them knows it."[1]

Yet the Zen experience can be passed on from teacher to pupil, and it has, in fact, been transmitted for many centuries by special methods proper to Zen. In a classic summary of four lines, Zen is described as

A special transmission outside the scriptures,
Not founded upon words and letters,
Pointing directly to the human mind,
Seeing into one's nature and attaining Buddhahood.

This technique of "direct pointing" constitutes the special flavor of Zen. It is typical of the Japanese mind which is more intuitive than intellectual and likes to give out facts as facts without much comment. The Zen masters were not given to verbosity and despised all theorizing and speculation. Thus they developed methods of pointing directly to the truth, with sudden and spontaneous actions or words, which expose the paradoxes of conceptual thinking and, like the *koans* I have already mentioned, are meant to stop the thought process to make the student ready for the mystical experience. This technique is well illustrated by the following examples of short conversations between master and disciple. In these conversations, which make up most of the Zen literature, the masters talk as little as possible and use their words to shift the disciples' attention from abstract thoughts to the concrete reality.

A monk, asking for instruction, said to Bodhidharma: "I have no peace of mind. Please pacify my mind."

"Bring your mind here before me," replied Bodhidharma, "and I will pacify it!"

"But when I seek my own mind," said the monk, "I cannot find it."

"There!" snapped Bodhidharma, "I have pacified your mind!"[2]

A monk told Joshu: "I have just entered the monastery. Please teach me."

Joshu asked: "Have you eaten your rice porridge?"

The monk replied: "I have eaten."
Joshu said: "Then you had better wash your bowl."[3]

These dialogues bring out another aspect which is characteristic of Zen. Enlightenment in Zen does not mean withdrawal from the world but means, on the contrary, active participation in everyday affairs. This viewpoint appealed very much to the Chinese mentality which attached great importance to a practical, productive life and to the idea of family perpetuation, and could not accept the monastic character of Indian Buddhism. The Chinese masters always stressed that *Ch'an*, or Zen, is our daily experience, the 'everyday mind' as Ma-tsu proclaimed. Their emphasis was on awakening in the midst of everyday affairs and they made it clear that they saw everyday life not only as the way to enlightenment, but as enlightenment itself.

In Zen, *satori* means the immediate experience of the Buddha nature of all things. First and foremost among these things are the objects, affairs and people involved in everyday life, so that while it emphasizes life's practicalities, Zen is nevertheless profoundly mystical. Living entirely in the present and giving full attention to everyday affairs, one who has attained *satori* experiences the wonder and mystery of life in every single act:

How wondrous this, how mysterious!
I carry fuel, I draw water.[4]

The perfection of Zen is thus to live one's everyday life naturally and spontaneously. When Po-chang was asked to define Zen, he said, "When hungry, eat, when tired, sleep." Although this sounds simple and obvious, like so much in Zen, it is in fact quite a difficult task. To regain the naturalness of our original nature requires long training and constitutes a great spiritual achievement. In the words of a famous Zen saying,

Before you study Zen, mountains are mountains and rivers are rivers; while you are studying Zen, mountains are no longer mountains and rivers are no longer rivers; but once you have had enlightenment, mountains are once again mountains and rivers again rivers.

Zen's emphasis on naturalness and spontaneity certainly shows its Taoist roots, but the basis for this emphasis is strictly Buddhistic. It is the belief in the perfection of our original nature, the realization that the process of enlightenment consists merely in becoming what we already are from the beginning. When the Zen master Po-chang was asked about seeking for the Buddha nature, he answered, "It's much like riding an ox in search of the ox."

There are two principal schools of Zen in Japan today which differ in their methods of teaching. The Rinzai or 'sudden' school uses the *koan* method, as discussed in a previous chapter, and gives prominence to periodic formal interviews with the master, called *sanzen*, during which the student is asked to present his view of the *koan* he is trying to solve. The solving of a *koan* involves long periods of intense concentration leading up to the sudden insight of *satori*. An experienced master knows when the student has reached the verge of sudden enlightenment and is able to shock him or her into the *satori* experience with unexpected acts, such as a blow with a stick or a loud yell.

The Soto or 'gradual' school avoids the shock methods of Rinzai and aims at the gradual maturing of the Zen student, "like the spring breeze which caresses the flower helping it to bloom."[5] It advocates 'quiet sitting' and the use of one's ordinary work as two forms of meditation.

Both the Soto and Rinzai schools attach the greatest importance to *zazen*, or sitting meditation, which is practiced in the Zen monasteries every day for many hours. The correct posture and breathing involved in this form of meditation is the first thing every student of Zen has to learn. In Rinzai Zen, *zazen* is used to prepare the intuitive mind for the handling of the *koan*, and the Soto school considers it as the most important means to help the student mature and evolve towards *satori*. More than that, it is seen as the actual realization of one's Buddha nature; body and mind being fused into a harmonious unity which needs no further improvement. As a Zen poem says,

> Sitting quietly, doing nothing,
> Spring comes, and the grass grows by itself.[6]

Since Zen asserts that enlightenment manifests itself in everyday affairs, it has had an enormous influence on all aspects of the traditional Japanese way of life. These include not only the arts of painting, calligraphy, garden design, etc., and the various crafts, but also ceremonial activities like serving tea or arranging flowers, and the martial arts of archery, swordsmanship, and *judo*. Each of these activities is known in Japan as a *do*, that is, a *tao* or 'way' toward enlightenment. They all explore various characteristics of the Zen experience and can be used to train the mind and to bring it in contact with the ultimate reality.

I have already mentioned the slow, ritualistic activities of *cha-no-yu*, the Japanese tea ceremony, the spontaneous movement of the hand required for calligraphy and painting, and the spirituality of *bushido*, the "way of the warrior." All these arts are expressions of the spontaneity, simplicity and total presence of mind characteristic of the Zen life. While they all require a perfection of technique, real mastery is only achieved when technique is transcended and the art becomes an "artless art" growing out of the unconscious.

We are fortunate to have a wonderful description of such an "artless art" in Eugen Herrigel's little book *Zen in the Art of Archery*. Herrigel spent more than five years with a celebrated Japanese master to learn his "mystical" art, and he gives us in his book a personal account of how he experienced Zen through archery. He describes how archery was presented to him as a religious ritual which is "danced" in spontaneous, effortless and purposeless movements. It took him many years of hard practice, which transformed his entire being, to learn how to draw the bow "spiritually," with a kind of effortless strength, and to release the string "without intention," letting the shot "fall from the archer like a ripe fruit." When he reached the height of perfection, bow, arrow, goal, and archer all melted into one another and he did not shoot, but "it" did it for him.

Herrigel's description of archery is one of the purest accounts of Zen because it does not talk about Zen at all.

$$\mathcal{L} = \bar{\psi} i\gamma^\mu D_\mu \psi - \tfrac{i}{2} g \bar{\psi} \gamma^\mu (D_\mu U)\psi - m\bar{\psi} U \psi + \tfrac{1}{16f^2} \mathrm{Tr}(D^\mu U D_\mu U^\dagger)$$

$$= \bar{N} i\gamma^\mu (D_\mu + M_\mu) N - m\bar{N}N + \tfrac{1}{16f^2} \mathrm{Tr}(D^\mu U D_\mu U^\dagger)$$

$$N = U^{1/2}\psi$$

$$D_\mu U = \partial_\mu U - 2igf\gamma_5 \rho\, \vec{\tau}(\vec{\varphi} \times \vec{\rho}_\mu) + ig\gamma_5 \sigma\, \vec{\tau}\vec{a}_\mu - 2gf\rho\, \vec{a}_\mu \vec{\varphi}$$

$$D_\mu N = \partial_\mu N - ig\left[\sigma \tfrac{\vec{\tau}}{2} - (1+g')\gamma_5 f\rho\, (\vec{\tau} \times \vec{\varphi}) + \tfrac{2f^2\rho^2}{1+\sigma} (\vec{\tau}\vec{\varphi})\vec{\varphi}\right] N \vec{\rho}_\mu$$

$$- ig\left[(1+g')\gamma_5 \sigma \tfrac{\vec{\tau}}{2} - f\rho(\vec{\tau} \times \vec{\varphi}) + (1+g')\gamma_5 \tfrac{2f^2\rho^2}{1+\sigma}(\vec{\tau}\vec{\varphi})\vec{\varphi}\right] N \vec{a}_\mu$$

$$M_\mu = U^{1/2}\partial_\mu U^{-1/2} + \tfrac{g'}{2}(U^{1/2}\partial_\mu U^\dagger) U^{-1/2}$$

$$U = \sigma(f^2\varphi^2) + 2if\gamma_5 \rho(f^2\varphi^2)\, \vec{\tau}\vec{\varphi}$$

$$\mathcal{L} = \bar{N} i\gamma^\mu (D_\mu + M_\mu) N - m\bar{N}N + \tfrac{1}{16f^2} \mathrm{Tr}(D^\mu U D_\mu U^\dagger)$$

$$- \tfrac{1}{4}\vec{R}^{\mu\nu}\vec{R}_{\mu\nu} - \tfrac{1}{4}\vec{A}^{\mu\nu}\vec{A}_{\mu\nu} + \tfrac{1}{2} m_\rho^2 (\vec{\rho}^{\,\mu}\vec{\rho}_\mu + \vec{a}^{\,\mu}\vec{a}_\mu)$$

$$\vec{R}_{\mu\nu} = \partial_\mu \vec{\rho}_\nu - \partial_\nu \vec{\rho}_\mu + g(\vec{\rho}_\mu \times \vec{\rho}_\nu) + g(\vec{a}_\mu \times \vec{a}_\nu)$$

$$\vec{A}_{\mu\nu} = \partial_\mu \vec{a}_\nu - \partial_\nu \vec{a}_\mu + g(\vec{a}_\mu \times \vec{\rho}_\nu) + g(\vec{\rho}_\mu \times \vec{a}_\nu)$$

$$\partial^\nu \vec{R}_{\mu\nu} = -g\vec{V}'_\mu + m_\rho^2 \vec{\rho}_\mu \;; \quad \vec{V}_\mu = \vec{V}'_\mu + \tfrac{1}{g}\partial^\nu \vec{R}_{\mu\nu}$$

$$\partial^\nu \vec{A}_{\mu\nu} = -g\vec{A}'_\mu + m_\rho^2 \vec{a}_\mu \;; \quad \vec{A}_\mu = \vec{A}'_\mu + \tfrac{1}{g}\partial^\nu \vec{A}_{\mu\nu}$$

$$[V_0^a(\vec{x},t), V_0^b(\vec{y},t)] = [A_0^a(\vec{x},t), A_0^b(\vec{y},t)] = i\varepsilon^{abc} V_0^c(\vec{x},t)\,\delta^3(\vec{x}-\vec{y})$$

$$[V_0^a(\vec{x},t), A_0^b(\vec{y},t)] = [A_0^a(\vec{x},t), V_0^b(\vec{y},t)] = i\varepsilon^{abc} A_0^c(\vec{x},t)\,\delta^3(\vec{x}-\vec{y})$$

$$[V_0^a(\vec{x},t), A_i^b(\vec{y},t)] = [A_0^a(\vec{x},t), V_i^b(\vec{y},t)] = i\varepsilon^{abc} A_i^c(\vec{x},t)\,\delta^3(\vec{x}-\vec{y})$$

$$[V_0^a(\vec{x},t), V_i^b(\vec{y},t)] = [A_0^a(\vec{x},t), A_i^b(\vec{y},t)] = i\varepsilon^{abc} V_i^c(\vec{x},t)\,\delta^3(\vec{x}-\vec{y})$$

III
THE PARALLELS

ऋतस्य यज्ञस्य धूर्षदं धुरि निर्वहणे सीदन्तं यज्ञनिर्वाहकमग्निं मित्रं न मित्रमिव समिधानः दूपीर्दीपयमानः ऋञ्जते। प्रसाधयति। ऋञ्जतिः प्रसाधनकर्मा। इन्धानः। सम्यग्दीप्यमानः। अक्रः। ज्वालासमिदादिभिराक्रान्तः। अन्यैरनाक्रान्तो वा। क्रमेश्छान्दसो डः। विदथेषु। यज्ञेषु वेदयत्सु स्तोत्रेषु निमित्तभूतेषु दीद्यत् स्वयं दीप्यमानोऽस्मदीयां धियं प्रज्ञां यागादिविषयां शुक्रवर्णां शुभ्रवर्णां निर्मलां ज्योतिष्टोमादि कर्म वा उद् यंसते। उद्योतयत्येव। यमेर्लेट्यडागमः। सिप्। उशब्दोऽवधारणे। धीरिति कर्मनाम। धीः शमीति तन्नामसु पाठात् ॥

मन्त्रः।

अप्रयुच्छन्नप्रयुच्छद्भिरग्ने
शिवेभिर्नः पायुभिः पाहि शग्मैः।
अदब्धेभिरदृपितेभिरिष्टे-
ऽनिमिषद्भिः परि पाहि नो जाः ॥ ८ ॥

पदपाठः।

अप्रऽयुच्छन्। अप्रयुच्छत्ऽभिः। अग्ने। शिवेभिः। नः। पायुऽभिः। पाहि। शग्मैः। अदब्धेभिः। अदृपितेभिः। इष्टे। अनिमिषत्ऽभिः। परि। पाहि। नः। जाः। ८ ॥

10
THE UNITY OF ALL THINGS

Although the spiritual traditions described in the last five chapters differ in many details, their view of the world is essentially the same. It is a view which is based on mystical experience—on a direct nonintellectual experience of reality—and this experience has a number of fundamental characteristics which are independent of the mystic's geographical, historical, or cultural background. A Hindu and a Taoist may stress different aspects of the experience; a Japanese Buddhist may interpret his or her experience in terms which are very different from those used by an Indian Buddhist; but the basic elements of the world-view which has been developed in all these traditions are the same. These elements also seem to be the fundamental features of the world view emerging from modern physics.

The most important characteristic of the Eastern world view—one could almost say the essence of it—is the awareness of the unity and mutual interrelation of all things and events, the experience of all phenomena in the world as manifestations of a basic oneness. All things are seen as interdependent and inseparable parts of this cosmic whole; as different manifestations of the same ultimate reality. The Eastern traditions constantly refer to this ultimate, indivisible reality which manifests itself in all things, and of which all things are parts. It is called *Brahman* in Hinduism, *Dharmakaya* in Buddhism, *Tao* in Taoism. Because it transcends all concepts and categories, Buddhists also call it *Tathata*, or Suchness:

> What is meant by the soul as suchness, is the oneness of the totality of all things, the great all-including whole.[1]

In ordinary life, we are not aware of this unity of all things, but divide the world into separate objects and events. This division is, of course, useful and necessary to cope with our everyday environment, but it is not a fundamental feature of reality. It is an abstraction devised by our discriminating and categorizing intellect. To believe that our abstract concepts of separate 'things' and 'events' are realities of nature is an illusion. Hindus and Buddhists tell us that this illusion is based on *avidya*, or ignorance, produced by a mind under the spell of *maya*. The principal aim of the Eastern mystical traditions is therefore to readjust the mind by centering and quietening it through meditation. The Sanskrit term for meditation—*samadhi*—means literally 'mental equilibrium.' It refers to the balanced and tranquil state of mind in which the basic unity of the universe is experienced:

> Entering into the *samadhi* of purity, [one obtains] all-penetrating insight that enables one to become conscious of the absolute oneness of the universe.[2]

The basic oneness of the universe is not only the central characteristic of the mystical experience, but is also one of the most important revelations of modern physics. It becomes apparent at the atomic level and manifests itself more and more as one penetrates deeper into matter, down into the realm of subatomic particles. The unity of all things and events will be a recurring theme throughout our comparison of modern physics and Eastern philosophy. As we study the various models of subatomic physics we shall see that they express again and again, in different ways, the same insight—that the constituents of matter and the basic phenomena involving them are all interconnected, interrelated and interdependent; that they cannot be understood as isolated entities, but only as integrated parts of the whole.

In this chapter, I shall discuss how the notion of the basic interconnectedness of nature arises in quantum theory, the theory of atomic phenomena, through a careful analysis of the process of observation.* Before entering this discussion, I have to return to the distinction between the mathematical framework of a theory and its verbal interpretation. The mathematical framework of quantum theory has passed countless successful tests and is now universally accepted as a consistent and accurate description of all atomic phenomena. The verbal interpretation, on the other hand—i.e., the metaphysics of quantum theory—is on far less solid ground. In fact, in more than forty years, physicists have not been able to provide a clear metaphysical model.

The following discussion is based on the so-called Copenhagen interpretation of quantum theory which was developed by Bohr and Heisenberg in the late 1920s and is still the most widely accepted model. In my discussion I shall follow the presentation given by Henry Stapp of the University of California[3] which concentrates on certain aspects of the theory and on a certain type of experimental situation that is frequently encountered in subatomic physics.** Stapp's presentation shows most clearly how quantum theory implies an essential interconnectedness of nature, and it also puts the theory in a framework that can readily be extended to the relativistic

* Although I have suppressed all the mathematics and simplified the analysis considerably, the following discussion may nevertheless appear to be rather dry and technical. It should perhaps be taken as "yogic" exercise which–like many exercises in the spiritual training of the Eastern traditions–may not be much fun, but may lead to a profound and beautiful insight into the essential nature of things.

** Other aspects of quantum theory will be discussed in subsequent chapters.

models of subatomic particles to be discussed later on.

The starting point of the Copenhagen interpretation is the division of the physical world into an observed system ("object") and an observing system. The observed system can be an atom, a subatomic particle, an atomic process, etc. The observing system consists of the experimental apparatus and will include one or several human observers. A serious difficulty now arises from the fact that the two systems are treated in different ways. The observing system is described in the terms of classical physics, but these terms cannot be used consistently for the description of the observed "object." We know that classical concepts are inadequate at the atomic level, yet we have to use them to describe our experiments and to state the results. There is no way we can escape this paradox. The technical language of classical physics is just a refinement of our everyday language and it is the only language we have to communicate our experimental results.

The observed systems are described in quantum theory in terms of probabilities. This means that we can never predict with certainty where a subatomic particle will be at a certain time, or how an atomic process will occur. All we can do is predict the odds. For example, most of the subatomic particles known today are unstable, that is, they disintegrate—or "decay"—into other particles after a certain time. It is not possible, however, to predict this time exactly. We can only predict the probability of decay after a certain time or, in other words, the average lifetime of a great number of particles of the same kind. The same applies to the "mode" of decay. In general, an unstable particle can decay into various combinations of other particles, and again we cannot predict which combination a particular particle will choose. All we can predict is that out of a large number of particles 60 percent, say, will decay in one way, 30 percent in another way, and 10 percent in a third way. It is clear that such statistical predictions need many measurements to be verified. Indeed, in the collision experiments of high-energy physics, tens of thousands of particle collisions are recorded and analyzed to determine the probability for a particular process.

It is important to realize that the statistical formulation

of the laws of atomic and subatomic physics does not reflect our ignorance of the physical situation, like the use of probabilities by insurance companies or gamblers. In quantum theory, we have come to recognize probability as a fundamental feature of the atomic reality which governs all processes, and even the existence of matter. Subatomic particles do not exist with certainty at definite places, but rather show "tendencies to exist," and atomic events do not occur with certainty at definite times and in definite ways, but rather show "tendencies to occur."

It is not possible, for example, to say with certainty where an electron will be in an atom at a certain time. Its position depends on the attractive force binding it to the atomic nucleus and on the influence of the other electrons in the atom. These conditions determine a probability pattern which represents the electron's tendencies to be in various regions of the atom. The picture

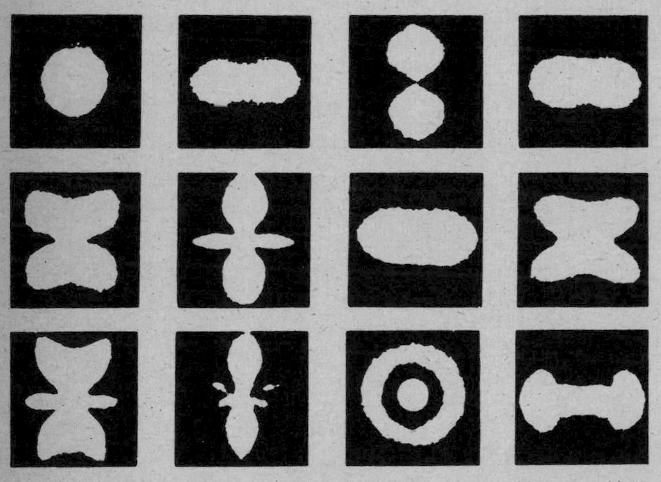

visual models of probability patterns

above shows some visual models of such probability patterns. The electron is likely to be found where the patterns are bright and unlikely to be present where they are dark. The important point is that the entire pattern represents the electron at a given time. Within the pattern, we cannot speak about the electron's position, but only about its tendencies to be in certain regions. In the mathematical formalism of quantum theory, these tendencies, or probabilities, are represented by the so-called probability function, a mathematical quantity which is related to the probabilities of finding the electron in various places at various times.

The contrast between the two kinds of description—classical terms for the experimental arrangement and probability functions for the observed objects—leads to deep metaphysical problems which have not yet been resolved. In practice, however, these problems are circumvented by describing the observing system in operational terms; that is, in terms of instructions which permit scientists to set up and carry out their experiments. In this way, the measuring devices and the scientists are effectively joined into one complex system which has no distinct, well-defined parts, and the experimental apparatus does not have to be described as an isolated physical entity.

For the further discussion of the process of observation it will be useful to take a definite example, and the simplest physical entity that can be used is a subatomic particle, such as the electron. If we want to observe and measure such a particle, we must first isolate it, or even create it, in a process which can be called the preparation process. Once the particle has been prepared for observation, its properties can be measured, and this constitutes the process of measurement. The situation can be represented symbolically as follows. A particle is prepared in the region A, travels from A to B, and is measured in the region B. In practice, both the preparation and the measurement of the particle may consist of a whole series of quite complicated processes. In the collision experiments of high-energy physics, for example, the preparation of the particles used as projectiles consists in sending them around a circular track and accelerating them until their energy is sufficiently high. This process takes place in the particle accelerator. When the desired energy is reached, they are made to leave the accelerator (A) and travel to the target area (B), where they collide with other particles. These collisions take place in a bubble chamber where the particles produce visible tracks which are photographed. The properties of the particles are then deduced from a mathematical analysis of their tracks; such an analysis can be quite complex and is often carried out with the help of computers. All these processes and activities constitute the act of measurement.

The important point in this analysis of observation is that the particle constitutes an intermediate system connecting the processes at A and B. It exists and has

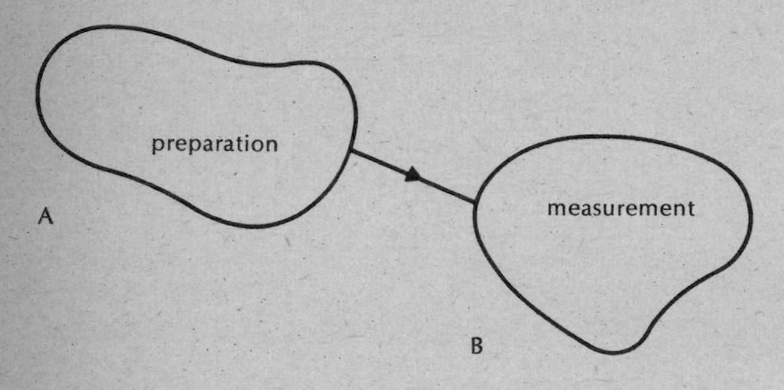

observation of a particle in atomic physics

meaning only in this context; not as an isolated entity, but as an interconnection between the processes of preparation and measurement. The properties of the particle cannot be defined independently of these processes. If the preparation or the measurement is modified, the properties of the particle will change, too.

On the other hand, the fact that we speak about "the particle," or any other observed system, shows that we have some independent physical entity in mind which is first prepared and then measured. The basic problem with observation in atomic physics is, then—in the words of Henry Stapp—that "the observed system is required to be isolated in order to be defined, yet interacting in order to be observed."[4] This problem is resolved in quantum theory in a pragmatic way by requiring that the observed system be free from the external disturbances caused by the process of observation during some interval between its preparation and subsequent measurement. Such a condition can be expected if the preparing and measuring devices are physically separated by a large distance, so that the observed object can travel from the region of preparation to the region of measurement.

How large, then, does this distance have to be? In principle, it must be infinite. In the framework of quantum theory, the concept of a distinct physical entity can be defined precisely only if this entity is infinitely far away from the agencies of observation. In practice, this is of course not possible; neither is it necessary. We have to remember, here, the basic attitude of modern science—that all its concepts and theories are approximate.* In the present case, this means that the concept of a distinct physical entity need not have a precise definition, but can be defined approximately. This is done in the following way.

The observed object is a manifestation of the interaction between the processes of preparation and measurement. This interaction is generally complex and involves various effects extending over different distances; it has various "ranges," as we say in physics. Now, if the dominant part of the interaction has a long range, the manifestation of this long-range effect will travel over a large distance. It will then be free from external disturbances and can be referred to as a distinct physical entity. In the framework of quantum theory, distinct physical entities are therefore idealizations which are meaningful only to the extent that the main part of the interaction has a long range. Such a situation can be defined mathematically in a precise way. Physically, it means that the measuring devices are placed so far apart that their main interaction occurs through the exchange of a particle or, in more complicated cases, of a network of particles. There will always be other effects present as well, but as long as the separation of the measuring devices is large enough these effects can be neglected. Only when the devices are not placed far enough apart will the short-range effects become dominant. In such a case, the whole macroscopic system forms a unified whole, and the notion of an observed object breaks down.

Quantum theory thus reveals an essential interconnectedness of the universe. It shows that we cannot decompose the world into independently existing smallest

* See p. 28.

units. As we penetrate into matter, we find that it is made of particles, but these are not the "basic building blocks" in the sense of Democritus and Newton. They are merely idealizations which are useful from a practical point of view, but have no fundamental significance. In the words of Niels Bohr, "Isolated material particles are abstractions, their properties being definable and observable only through their interaction with other systems."[5]

The Copenhagen interpretation of quantum theory is not universally accepted. There are several counterproposals, and the philosophical problems involved are far from being settled. The universal interconnectedness of things and events, however, seems to be a fundamental feature of the atomic reality which does not depend on a particular interpretation of the mathematical theory. The following passage from a recent article by David Bohm, one of the main opponents of the Copenhagen interpretation, confirms this fact most eloquently:

> One is led to a new notion of unbroken wholeness which denies the classical idea of analyzability of the world into separately and independently existing parts. . . . We have reversed the usual classical notion that the independent "elementary parts" of the world are the fundamental reality, and that the various systems are merely particular contingent forms and arrangements of these parts. Rather, we say that inseparable quantum interconnectedness of the whole universe is the fundamental reality, and that relatively independently behaving parts are merely particular and contingent forms within this whole.[6]

At the atomic level, then, the solid material objects of classical physics dissolve into patterns of probabilities, and these patterns do not represent probabilities of things, but rather probabilities of interconnections. Quantum theory forces us to see the universe not as a collection of physical objects, but rather as a complicated web of relations between the various parts of a unified whole. This, however, is the way in which Eastern mystics have experienced the world, and some of them have expressed their experience in words which are almost identical with those used by atomic physicists. Here are two examples:

> The material object becomes . . . something different from what we now see, not a separate object on the background or in the environment of the rest of nature but an indivisible part and even in a subtle way an expression of the unity of all that we see.[7]

> Things derive their being and nature by mutual dependence and are nothing in themselves.[8]

If these statements could be taken as an account of how nature appears in atomic physics, the following two statements from atomic physicists could, in turn, be read as a description of the mystical experience of nature:

> An elementary particle is not an independently existing unanalyzable entity. It is, in essence, a set of relationships that reach outward to other things.[9]

> The world thus appears as a complicated tissue of events, in which connections of different kinds alternate or overlap or combine and thereby determine the texture of the whole.[10]

The picture of an interconnected cosmic web which emerges from modern atomic physics has been used extensively in the East to convey the mystical experience of nature. For the Hindus, *Brahman* is the unifying thread in the cosmic web, the ultimate ground of all being:

> He on whom the sky, the earth, and the atmosphere
> Are woven, and the wind, together with all life-breaths,
> Him alone know as the one Soul.[11]

In Buddhism, the image of the cosmic web plays an even greater role. The core of the *Avatamsaka Sutra*, one of the main scriptures of Mahayana Buddhism,* is the description of the world as a perfect network of mutual relations where all things and events interact with each other in

* See p. 89.

an infinitely complicated way. Mahayana Buddhists have developed many parables and similes to illustrate this universal interrelatedness, some of which will be discussed later on, in connection with the relativistic version of the 'web philosophy' in modern physics. The cosmic web, finally, plays a central role in Tantric Buddhism, a branch of the Mahayana which originated in India around the third century A.D. and constitutes today the main school of Tibetan Buddhism. The scriptures of this school are called the *Tantras*, a word whose Sanskrit root means "to weave" and which refers to the interwovenness and interdependence of all things and events.

In Eastern mysticism, this universal interwovenness always includes the human observer and his or her consciousness, and this is also true in atomic physics. At the atomic level, "objects" can be understood only in terms of the interaction between the processes of preparation and measurement. The end of this chain of processes lies always in the consciousness of the human observer. Measurements are interactions which create "sensations" in our consciousness—for example, the visual sensation of a flash of light, or of a dark spot on a photographic plate—and the laws of atomic physics tell us with what probability an atomic object will give rise to a certain sensation if we let it interact with us. "Natural science," says Heisenberg, "does not simply describe and explain nature; it is part of the interplay between nature and ourselves."[12]

The crucial feature of atomic physics is that the human observer is not only necessary to observe the properties of an object, but is necessary even to define these properties. In atomic physics, we cannot talk about the properties of an object as such. They are meaningful only in the context of the object's interaction with the observer. In the words of Heisenberg, "What we observe is not nature itself, but nature exposed to our method of questioning."[13] The observer decides how he is going to set up the measurement and this arrangement will determine, to some extent, the properties of the observed object. If the experimental arrangement is modified, the properties of the observed object will change in turn.

This can be illustrated with the simple case of a subatomic particle. When observing such a particle, one may choose to measure—among other quantities—the particle's position and its momentum (a quantity defined as the particle's mass times its velocity). We shall see in the next chapter that an important law of quantum theory—Heisenberg's uncertainty principle—says that these two quantities can never be measured simultaneously with precision. We can either obtain a precise knowledge about the particle's position and remain completely ignorant about its momentum (and thus about its velocity), or vice versa; or we can have a rough and imprecise knowledge about both quantities. The important point now is that this limitation has nothing to do with the imperfection of our measuring techniques. It is a principle limitation which is inherent in the atomic reality. If we decide to measure the particle's position precisely, the particle simply *does not have* a well-defined momentum, and if we decide to measure the momentum, it does not have a well-defined position.

In atomic physics, then, the scientist cannot play the role of a detached objective observer, but becomes involved in the world he observes to the extent that he influences the properties of the observed objects. John Wheeler sees this involvement of the observer as the most important feature of quantum theory and he has therefore suggested replacing the word "observer" by the word "participator." In Wheeler's own words,

> Nothing is more important about the quantum principle than this, that it destroys the concept of the world as "sitting out there," with the observer safely separated from it by a 20-centimeter slab of plate glass. Even to observe so minuscule an object as an electron, he must shatter the glass. He must reach in. He must install his chosen measuring equipment. It is up to him to decide whether he shall measure position or momentum. To install the equipment to measure the one prevents and excludes his installing the equipment to measure the other. Moreover, the measurement changes the state of the electron. The universe will never afterward be the same. To describe what has happened, one has to cross out that old word "observer" and put in its place the new word "participator." In

> some strange sense, the universe is a participatory universe.[14]

The idea of "participation instead of observation" has been formulated in modern physics only recently, but it is an idea which is well known to any student of mysticism. Mystical knowledge can never be obtained just by observation, but only by full participation with one's whole being. The notion of the participator is thus crucial to the Eastern world-view, and the Eastern mystics have pushed this notion to the extreme, to a point where observer and observed, subject and object, are not only inseparable but also become indistinguishable. The mystics are not satisfied with a situation analogous to atomic physics, where the observer and the observed cannot be separated, but can still be distinguished. They go much further, and in deep meditation they arrive at a point where the distinction between observer and observed breaks down completely, where subject and object fuse into a unified undifferentiated whole. Thus the *Upanishads* say:

> Where there is a duality, as it were, there one sees another; there one smells another; there one tastes another. . . . But where everything has become just one's own self, then whereby and whom would one see? Then whereby and whom would one smell? Then whereby and whom would one taste?[15]

This, then, is the final apprehension of the unity of all things. It is reached—so the mystics tell us—in a state of consciousness where one's individuality dissolves into an undifferentiated oneness, where the world of the senses is transcended and the notion of "things" is left behind. In the words of Chuang Tzu:

> My connection with the body and its parts is dissolved. My perceptive organs are discarded. Thus leaving my material form and bidding farewell to my knowledge, I become one with the Great Pervader. This I call sitting and forgetting all things.[16]

Modern physics, of course, works in a very different framework and cannot go that far in the experience of the unity of all things. But it has made a great step toward the world-view of the Eastern mystics in atomic theory. Quantum theory has abolished the notion of fundamentally separated objects, has introduced the concept of the participator to replace that of the observer, and may even find it necessary to include the human consciousness in its description of the world.* It has come to see the universe as an interconnected web of physical and mental relations whose parts are defined only through their connections to the whole. To summarize the world-view emerging from atomic physics, the words of a Tantric Buddhist, Lama Anagarika Govinda, seem to be perfectly apropos:

> The Buddhist does not believe in an independent or separately existing external world, into whose dynamic forces he could insert himself. The external world and his inner world are for him only two sides of the same fabric, in which the threads of all forces and of all events, of all forms of consciousness and of their objects, are woven into an inseparable net of endless, mutually conditioned relations.[17]

* This point will be discussed further in Chapter 18.

11
BEYOND THE WORLD OF OPPOSITES

When the Eastern mystics tell us that they experience all things and events as manifestations of a basic oneness, this does not mean that they pronounce all things to be equal. They recognize the individuality of things, but at the same time they are aware that all differences and contrasts are relative within an all-embracing unity. Since in our normal state of consciousness, this unity of all contrasts—and especially the unity of opposites—is extremely hard to accept, it constitutes one of the most puzzling features of Eastern philosophy. It is, however, an insight which lies at the very root of the Eastern world-view.

Opposites are abstract concepts belonging to the realm of thought, and as such they are relative. By the very act of focusing our attention on any one concept, we create its opposite. As Lao Tzu says, "When all in the world understand beauty to be beautiful, then ugliness exists; when all understand goodness to be good, then evil exists."[1] The mystic transcends this realm of intellectual concepts, and in transcending it he becomes aware of the relativity and polar relationship of all opposites. He realizes that good and bad, pleasure and pain, life and death, are not absolute experiences belonging to different categories, but are merely two sides of the same reality; extreme parts of a single whole. The awareness that all opposites are polar, and thus a unity, is seen as one of the highest aims of man in the spiritual traditions of the East. "Be in truth eternal, beyond earthly opposites!" is Krishna's advice in the *Bhagavad Gita*, and the same advice is given to the followers of Buddhism. Thus D. T. Suzuki writes:

> The fundamental idea of Buddhism is to pass beyond the world of opposites, a world built up by intellectual distinctions and emotional defilements, and to realize the spiritual world of nondistinction, which involves achieving an absolute point of view.[2]

The whole of Buddhist teaching—and in fact the whole of Eastern mysticism—revolves about this absolute point of view which is reached in the world of *acintya*, or 'no-thought,' where the unity of all opposites becomes a vivid experience. In the words of a Zen poem,

> At dusk the cock announces dawn;
> At midnight, the bright sun.[3]

The notion that all opposites are polar—that light and dark, winning and losing, good and evil, are merely different aspects of the same phenomenon—is one of the basic principles of the Eastern way of life. Since all opposites are interdependent, their conflict can never result in the total victory of one side, but will always be a manifestation of the interplay between the two sides. In the East, a virtuous person is therefore not one who undertakes the impossible task of striving for the good and eliminating the bad, but rather one who is able to maintain a dynamic balance between good and bad.

This notion of dynamic balance is essential to the way in which the unity of opposites is experienced in Eastern mysticism. It is never a static identity, but always a dynamic interplay between two extremes. This point has been emphasized most extensively by the Chinese sages in their symbolism of the archetypal poles *yin* and *yang*. They called the unity lying behind *yin* and *yang* the *Tao* and saw it as a process which brings about their interplay: "That which lets now the dark, now the light appear is *Tao*."[4]

The dynamic unity of polar opposites can be illustrated

with the simple example of a circular motion and its projection. Suppose you have a ball going around a circle. If this movement is projected onto a screen, it becomes an oscillation between two extreme points. (To keep the analogy with Chinese thought, I have written TAO in the circle and have marked the extreme points of the oscillation with YIN and YANG.) The ball goes round the circle with constant speed, but in the projection it slows down as it reaches the edge, turns around, and then accelerates again only to slow down once more

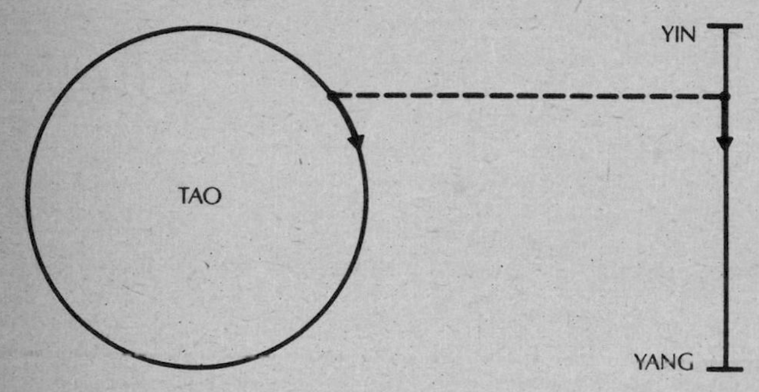

dynamic unity of polar opposites

—and so on, in endless cycles. In any projection of that kind, the circular movement will appear as an oscillation between two opposite points, but in the movement itself the opposites are unified and transcended. This image of a dynamic unification of opposites was indeed very much in the minds of the Chinese thinkers, as can be seen from the passage in the *Chuang-tzu* quoted previously:*

> That the "that" and the "this" cease to be opposites is the very essence of *Tao*. Only this essence, an axis as it were, is the center of the circle responding to the endless changes.

One of the principal polarities in life is the one between the male and female sides of human nature. As with the polarity of good and bad, or of life and death, we tend

* See p. 102.

to feel uncomfortable with the male/female polarity in ourselves, and therefore we bring one or the other side into prominence. Western society has traditionally favored the male side rather than the female. Instead of recognizing that the personality of each man and of each woman is the result of an interplay between female and male elements, it has established a static order where all men are supposed to be masculine and all women feminine, and it has given men the leading roles and most of society's privileges. This attitude has resulted in an overemphasis of all the *yang*—or male—aspects of human nature: activity, rational thinking, competition, aggressiveness, and so on. The *yin*—or female—modes of consciousness, which can be described by words like intuitive, religious, mystical, occult, or psychic, have constantly been suppressed in our male-oriented society.

In Eastern mysticism, these female modes are developed and a unity between the two aspects of human nature is sought. A fully realized human being is one who, in the words of Lao Tzu, "knows the masculine and yet keeps to the feminine." In many Eastern traditions the dynamic balance between the male and female modes of consciousness is the principal aim of meditation, and is often illustrated in works of art. A superb sculpture of Shiva in the Hindu temple of Elephanta shows three faces of the god: on the right, his male profile displaying virility and willpower; on the left, his female aspect—gentle, charming, seductive—and in the center the sublime union of the two aspects in the magnificent head of Shiva Mahesvara, the Great Lord, radiating serene tranquillity and transcendental aloofness. In the same temple, Shiva is also represented in androgynous form—half-male, half-female—the flowing movement of the god's body and the serene detachment of his/her face symbolizing, again, the dynamic unification of the male and female.

In Tantric Buddhism, the male/female polarity is often illustrated with the help of sexual symbols. Intuitive wisdom is seen as the passive, female quality of human nature, love and compassion as the active, male quality; and the union of both in the process of enlightenment is represented by ecstatic sexual embraces of male and female deities. The Eastern mystics affirm that such a union of one's male and female modes can be experienced only on a higher plane of consciousness where the realm of thought and language is transcended and all opposites appear as a dynamic unity.

I have already asserted that a similar plane has been reached in modern physics. The exploration of the subatomic world has revealed a reality which repeatedly transcends language and reasoning, and the unification of concepts which had hitherto seemed opposite and irreconcilable turns out to be one of the most startling features of this new reality. These seemingly irreconcilable concepts are generally not the ones the Eastern mystics are concerned with—although sometimes they are—but their unification at a nonordinary level of reality provides a parallel to Eastern mysticism. Modern physicists should therefore be able to gain insights into some of the central teachings of the Far East by relating them to experiences in their own field. A small but growing number of young physicists have indeed found this a most valuable and stimulating approach to Eastern mysticism.

Examples of the unification of opposite concepts in modern physics can be found at the subatomic level, where particles are both destructible and indestructible; where matter is both continuous and discontinuous, and force and matter are but different aspects of the same phenomenon. In all these examples, which will be discussed extensively in subsequent chapters, it turns out that the framework of opposite concepts, derived from our everyday experience, is too narrow for the world of subatomic particles. Relativity theory is crucial for the description of this world, and in the "relativistic" framework the classical concepts are transcended by going to a higher dimension, the four-dimensional space-time. Space and time themselves are two concepts which had seemed entirely different, but have been unified in relativistic physics. This fundamental unity is the basis of the unification of the opposite concepts mentioned above. Like the unity of opposites experienced by the mystics, it takes place on a "higher plane," i.e., in a higher dimension, and like that experienced by the mystics it is a dynamic unity, because the relativistic space-time reality is an intrinsically dynamic reality where objects are also processes and all forms are dynamic patterns.

To experience the unification of seemingly separate entities in a higher dimension we do not need relativity theory. It can also be experienced by going from one to two dimensions, or from two to three. In the example of a circular motion and its projection given on p. 132 the opposite poles of the oscillation in one dimension (along a line) are unified in the circular movement in two dimensions (in one plane). The drawing below

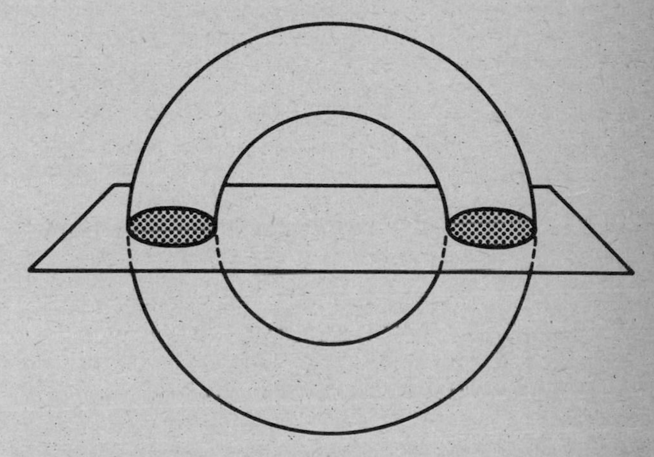

represents another example, involving a transition from two to three dimensions. It shows a "doughnut" ring cut horizontally by a plane. In the two dimensions of that plane, the surfaces of the cut appear as two completely separate discs, but in three dimensions they are recognized as being parts of one and the same object. A similar unification of entities which seem separate and irreconcilable is achieved in relativity theory by going from three to four dimensions. The four-dimensional world of relativistic physics is the world where force and matter are unified; where matter can appear as discontinuous particles or as a continuous field. In these cases, however, we can no longer visualize the unity very well. Physicists can "experience" the four-dimensional space-time world through the abstract mathematical formalism of their theories, but their visual imagination—like everybody else's—is limited to the three-dimensional world of the senses. Our language and thought patterns have evolved in this three-dimensional world, and therefore we find it extremely hard to deal with the four-dimensional reality of relativistic physics.

Eastern mystics, on the other hand, seem to be able to experience a higher-dimensional reality directly and concretely. In the state of deep meditation, they can transcend the three-dimensional world of everyday life and experience a totally different reality, where all opposites are unified into an organic whole. When the mystics try to express this experience in words, they are faced with the same problems as the physicists trying to interpret the multidimensional reality of relativistic physics. In the words of Lama Govinda:

> An experience of higher dimensionality is achieved by integration of experiences of different centers and levels of consciousness. Hence the indescribability of certain experiences of meditation on the plane of three-dimensional consciousness and within a system of logic which reduces the possibilities of expression by imposing further limits upon the process of thinking.[5]

The four-dimensional world of relativity theory is not the only example in modern physics where seemingly contradictory and irreconcilable concepts are seen to be nothing more than different aspects of the same reality. Perhaps the most famous case of such a unification of contradictory concepts is that of the concepts of particles and waves in atomic physics.

At the atomic level, matter has a dual aspect: it appears as particles and as waves. Which aspect it shows depends on the situation. In some situations the particle aspect is dominant, in others the particles behave more like waves; and this dual nature is also exhibited by light and all other electromagnetic radiation. Light, for example, is emitted and absorbed in the form of "quanta," or photons, but when these particles of light travel through space, they appear as vibrating electric and magnetic fields which show all the characteristic behavior of waves. Electrons are normally considered to be particles, and yet when a beam of these particles is sent through a small slit, it is diffracted just like a beam of light—in other words, electrons, too, behave like waves.

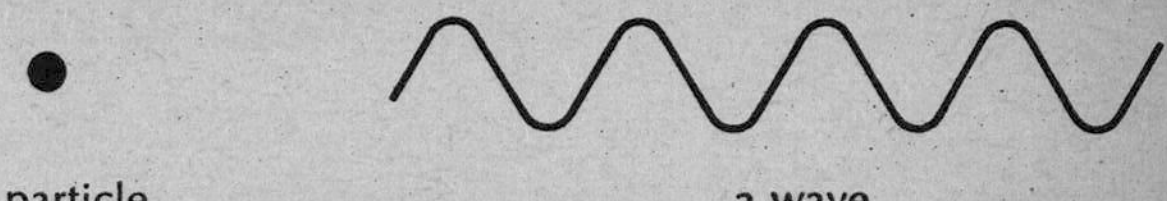

a particle a wave

This dual aspect of matter and radiation is indeed most startling and gave rise to many of the "quantum *koans*" which led to the formulation of quantum theory. The picture of a wave which is always spread out in space is fundamentally different from the particle picture which implies a sharp location. It has taken physicists a long time to accept the fact that matter manifests itself in ways which seem to be mutually exclusive; that particles are also waves, waves also particles.

Looking at the two pictures, a lay person might be tempted to think that the contradiction can be resolved by saying that the picture on the right-hand side simply represents a particle moving in a wave pattern. This argument, however, rests on a misunderstanding of the nature of waves. Particles moving in wave patterns do not exist in nature. In a water wave, for example, the water particles do not move along with the wave but move in circles as the wave passes by. Similarly, the air particles in a sound wave merely oscillate back and forth, but do not propagate along with the wave. What is transported along the wave is the disturbance causing the wave phenomenon, but not any material particle. In quantum theory, therefore, we do not speak about a particle's trajectory when we say that the particle is also

direction of wave

a water wave

a wave. What we mean is that the wave pattern as a whole is a manifestation of the particle. The picture of traveling waves is thus totally different from that of traveling particles; as different—in the words of Victor Weisskopf—"as the notion of waves on a lake from that of a school of fish swimming in the same direction."[6]

The phenomenon of waves is encountered in many different contexts throughout physics and can be described with the same mathematical formalism whenever it occurs. The same mathematical forms are used to describe a light wave, a vibrating guitar string, a sound wave, or a water wave. In quantum theory, these forms are used again to describe the waves associated with particles. This time, however, the waves are much more abstract. They are closely related to the statistical nature of quantum theory, i.e., to the fact that atomic phenomena can only be described in terms of probabilities. The information about the probabilities for a particle is contained in a quantity called the probability function, and the mathematical form of this quantity is that of a wave, that is to say, it is similar to the forms used for the description of other types of waves. The waves associated with particles, however, are not "real" three-dimensional waves, like water waves or sound waves, but are "probability waves"; abstract mathematical quantities which are related to the probabilities of finding the particles in various places and with various properties.

The introduction of probability waves, in a sense, resolves the paradox of particles being waves by putting it in a totally new context; but at the same time it leads to another pair of opposite concepts which is even more fundamental, that of existence and nonexistence. This pair of opposites, too, is transcended by the atomic reality. We can never say that an atomic particle exists at a certain place, nor can we say that it does not exist. Being a probability pattern, the particle has tendencies to exist in various places and thus manifests a strange kind of physical reality between existence and nonexistence. We cannot, therefore, describe the state of the particle in terms of fixed opposite concepts. The particle is not present at a definite place, nor is it absent. It does not change its position, nor does it remain at rest. What changes is the probability pattern, and thus the tendencies of the particle to exist in certain places. In the words of Robert Oppenheimer,

> If we ask, for instance, whether the position of the electron remains the same, we must say "no"; if we ask whether the electron's position changes with time, we must say "no"; if we ask whether the electron is at rest, we must say "no"; if we ask whether it is in motion, we must say "no."[7]

The reality of the atomic physicist, like the reality of the Eastern mystic, transcends the narrow framework of opposite concepts. Oppenheimer's words thus seem to echo the words of the *Upanishads*,

> It moves. It moves not.
> It is far, and It is near.
> It is within all this,
> And It is outside of all this.[8]

Force and matter, particles and waves, motion and rest, existence and nonexistence—these are some of the opposite or contradictory concepts which are transcended in modern physics. Of all these opposite pairs, the last seems to be the most fundamental, and yet, in atomic physics we have to go even beyond the concepts of existence and nonexistence. This is the feature of quantum theory which is most difficult to accept and which lies at the heart of the continuing discussion about its interpretation. At the same time, the transcending of the concepts of existence and nonexistence is also one of the most puzzling aspects of Eastern mysticism. Like the atomic physicists, the Eastern mystics deal with a reality which lies beyond existence and nonexistence, and they frequently emphasize this important fact. Thus Ashvaghosha:

> Suchness is neither that which is existence, nor that which is nonexistence, nor that which is at once existence and nonexistence, nor that which is not at once existence and nonexistence.[9]

Faced with a reality which lies beyond opposite concepts, physicists and mystics have to adopt a special way of thinking, where the mind is not fixed in the rigid framework of classical logic, but keeps moving and changing its viewpoint. In atomic physics, for example, we are now used to applying both the particle and the wave concept in our description of matter. We have learned how to play with the two pictures, switching from one to the other and back, in order to cope with the atomic reality. This is precisely the way in which the Eastern mystics think when they try to interpret their experience of a reality beyond opposites. In the words of Lama Govinda, "The Eastern way of thinking rather consists in a circling round the object of contemplation . . . a many-sided, i.e., multidimensional impression formed from the superimposition of single impressions from different points of view."[10]

To see how one can switch back and forth between the particle picture and the wave picture in atomic physics, let us examine the concepts of waves and particles in more detail. A wave is a vibrational pattern in space and time. We can look at it at a definite instant of time and will then see a periodic pattern in space, as in the following example. This pattern is characterized by an amplitude A, the extension of the vibration, and a wavelength L, the distance between two successive crests.

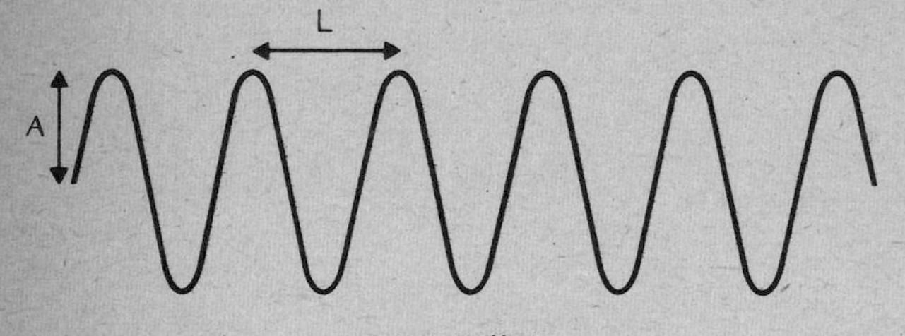

a wave pattern

Alternatively, we can look at the motion of a definite point of the wave and will then see an oscillation characterized by a certain frequency, the number of times the point oscillates back and forth every second. Now let us turn to the particle picture. According to classical ideas, a particle has a well-defined position at any time, and its state of motion can be described in terms of its velocity and its energy of motion. Particles moving with a high velocity also have a high energy.

Physicists, in fact, hardly use "velocity" to describe the particle's state of motion, but rather use a quantity called "momentum" which is defined as the particle's mass times its velocity.

Quantum theory, now, associates the properties of a probability wave with the properties of the corresponding particle by relating the amplitude of the wave at a certain place to the probability of finding the particle at that place. Where the amplitude is large we are likely to find the particle if we look for it, where it is small, unlikely. The wave train pictured on p. 140, for example, has the same amplitude throughout its length, and the particle can therefore be found anywhere along the wave with the same likelihood.*

The information about the particle's state of motion is contained in the wavelength and frequency of the wave. The wavelength is inversely proportional to the momentum of the particle, which means that a wave with a small wavelength corresponds to a particle moving with a high momentum (and thus with a high velocity). The frequency of the wave is proportional to the particle's energy; a wave with a high frequency means that the particle has a high energy. In the case of light, for example, violet light has a high frequency and a short wavelength and consists therefore of photons of high energy and high momentum; whereas red light has a low frequency and a long wavelength, corresponding to photons of low energy and momentum.

A wave which is spread out like the one in our example does not tell us much about the position of the corresponding particle. It can be found anywhere along the wave with the same likelihood. Very often, however, we deal with situations where the particle's position is known to some extent, as for example in the description of an electron in an atom. In such a case, the probabilities of finding the particle in various places must be confined to a certain region. Outside this region they

* In this example, one must not think that the particle is more likely to be found where the wave has crests than in the places where it has troughs. The static wave pattern in the picture is merely a "snapshot" of a continual vibration during which every point along the wave reaches the top of a crest in periodic intervals.

must be zero. This can be achieved by a wave pattern like the one in the following diagram which corresponds to a particle confined to the region X. Such a pattern is called a wave packet.* It is composed of several wave

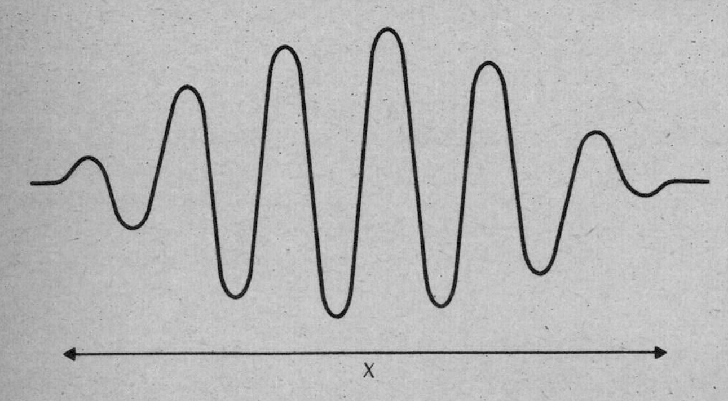

a wave packet corresponding to a particle
located somewhere in the region X

trains with various wavelengths which interfere with each other destructively** outside the region X, so that the total amplitude—and thus the probability of finding the particle there—is zero, whereas they build up the pattern inside X. This pattern shows that the particle is located somewhere inside the region X, but it does not allow us to localize it any further. For points inside the region we can only give the probabilities for the presence of the particle. (The particle is most likely to be present in the center where the probability amplitudes are large, and less likely near the ends of the wave packet where the amplitudes are small.) The length of the wave packet represents therefore the uncertainty in the location of the particle.

The important property of such a wave packet now is that it has no definite wavelength, i.e., the distances between two successive crests are not equal throughout

* For simplicity, we deal here only with one dimension of space, i.e., with the position of the particle somewhere along a line. The probability patterns shown on p. 120 are two-dimensional examples corresponding to more complicated wave packets.

** See p. 34.

the pattern. There is a spread in wavelength the amount of which depends on the length of the wave packet: the shorter the wave packet, the larger the spread in wavelength. This has nothing to do with quantum theory, but simply follows from the properties of waves. Wave packets do not have a definite wavelength. Quantum theory comes into play when we associate the wavelength with the momentum of the corresponding particle. If the wave packet does not have a well-defined wavelength, the particle does not have a well-defined momentum. This means that there is not only an uncertainty in the particle's position, corresponding to the length of the wave packet, but also an uncertainty in its momentum, caused by the spread in wavelength. The two uncertainties are interrelated, because the spread in wavelength (i.e., the uncertainty of momentum) depends on the length of the wave packet (i.e., on the uncertainty of position). If we want to localize the particle more precisely, that is, if we want to confine its wave packet to a smaller region, this will result in an increase in the spread in wavelength and thus in an increase in the uncertainty of the particle's momentum.

The precise mathematical form of this relation between the uncertainties of position and momentum of a particle is known as Heisenberg's uncertainty relation, or uncertainty principle. It means that, in the subatomic world, we can never know both the position and momentum of a particle with great accuracy. The better we know the position, the hazier will its momentum be and vice versa. We can decide to undertake a precise measurement of either of the two quantities; but then we will have to remain completely ignorant about the other one. It is important to realize, as was pointed out in the previous chapter, that this limitation is not caused by the imperfection of our measuring techniques, but is a limitation of principle. If we decide to measure the particle's position precisely, the particle simply does not have a well-defined momentum, and vice versa.

The relation between the uncertainties of a particle's position and momentum is not the only form of the uncertainty principle. Similar relations hold between other quantities, for example between the time an atomic event takes and the energy it involves. This can be seen quite easily by picturing our wave packet not as a pattern in space, but as a vibrational pattern in time. As the particle passes a particular point of observation, the vibrations of the wave pattern at that point will start with small amplitudes which will increase and then decrease again until finally the vibration will stop altogether. The time it takes to go through this pattern represents the time during which the particle passes our point of observation. We can say that the passage occurs within this time span, but we cannot localize it any further. The duration of the vibration pattern represents therefore the uncertainty in the temporal location of the event.

Now, as the spatial pattern of the wave packet does not have a well-defined wavelength, the corresponding vibrational pattern in time does not have a well-defined frequency. The spread in frequency depends on the duration of the vibrational pattern, and since quantum theory associates the frequency of the wave with the energy of the particle, the spread in the pattern's frequency corresponds to an uncertainty in the particle's energy. The uncertainty in the location of an event in time thus becomes related to an uncertainty in energy in the same way as the uncertainty of a particle's location in space is related to an uncertainty in momentum. This means that we can never know both the time at which an event occurs and the energy involved in it with great accuracy. Events occurring inside a short time span involve a large uncertainty in energy; events involving a precise amount of energy can be localized only within a long period of time.

The fundamental importance of the uncertainty principle is that it expresses the limitations of our classical concepts in a precise mathematical form. As described previously, the subatomic world appears as a web of relations between the various parts of a unified whole. Our classical notions, derived from our ordinary macroscopic experience, are not fully adequate to describe this world. To begin with, the concept of a distinct physical entity, like a particle, is an idealization which has no fundamental significance. It can only be defined in terms of its connections to the whole, and these connections are of a statistical nature—probabilities rather than certainties. When we describe the properties of such an entity in terms of classical concepts—like position, energy, momentum, etc.—we find that there are pairs of concepts which are interrelated and cannot be defined simultaneously in a precise way. The more we impose one concept on the physical "object," the more the other concept becomes uncertain, and the precise relation between the two is given by the uncertainty principle.

For a better understanding of this relation between pairs of classical concepts, Niels Bohr has introduced the notion of complementarity. He considered the particle picture and the wave picture as two complementary descriptions of the same reality, each of them being only partly correct and having a limited range of application. Each picture is needed to give a full description of the atomic reality, and both are to be applied within the limitations given by the uncertainty principle.

This notion of complementarity has become an essential part of the way physicists think about nature, and Bohr has often suggested that it might be a useful concept also outside the field of physics; in fact, the notion of complementarity proved to be extremely useful 2,500 years ago. It played an essential role in ancient Chinese thought which was based on the insight that opposite concepts stand in a polar—or complementary—relationship to each other. The Chinese sages represented this complementarity of opposites by the archetypal poles *yin* and *yang* and saw their dynamic interplay as the essence of all natural phenomena and all human situations.

Niels Bohr was well aware of the parallel between his concept of complementarity and Chinese thought. When he visited China in 1937, at a time when his interpretation of quantum theory had already been fully elaborated, he was deeply impressed by the ancient Chinese notion of polar opposites, and from that time he maintained an interest in Eastern culture. Ten years later, Bohr was knighted as an acknowledgment of his outstanding achievements in science and important contributions to Danish cultural life; and when he had to choose a suitable motif for his coat-of-arms, his choice fell on the Chinese symbol of *t'ai-chi* representing the complementary relationship of the archetypal opposites *yin* and *yang*. In choosing this symbol for his coat-of-arms together with the inscription *Contraria sunt complementa* (Opposites are complementary), Niels Bohr acknowledged the profound harmony between ancient Eastern wisdom and modern Western science.

12
SPACE-TIME

$$R_{\mu\nu} - \frac{1}{2} g_{\mu\nu} R = \kappa T_{\mu\nu}$$

Modern physics has confirmed most dramatically one of the basic ideas of Eastern mysticism; that all the concepts we use to describe nature are limited, that they are not features of reality, as we tend to believe, but creations of the mind; parts of the map, not of the territory. Whenever we expand the realm of our experience, the limitations of our rational mind become apparent and we have to modify, or even abandon, some of our concepts.

Our notions of space and time figure prominently on our map of reality. They serve to order things and events in our environment and are therefore of paramount importance not only in our everyday life, but also in our attempts to understand nature through science and philosophy. There is no law of physics which does not require the concepts of space and time for its formulation. The profound modification of these basic concepts brought about by relativity theory was therefore one of the greatest revolutions in the history of science.

Classical physics was based on the notion both of an absolute, three-dimensional space, independent of the material objects it contains, and obeying the laws of Euclidean geometry, and of time as a separate dimension which again is absolute and flows at an even rate, independent of the material world. In the West, these notions of space and time were so deeply rooted in the minds of philosophers and scientists that they were taken as true and unquestioned properties of nature.

The belief that geometry is inherent in nature, rather than part of the framework we use to describe nature, has its origin in Greek thought. Demonstrative geometry was the central feature of Greek mathematics and had a profound influence on Greek philosophy. Its method of starting from unquestioned axioms, and deriving theorems from these by deductive reasoning, became characteristic of Greek philosophical thought; geometry was therefore at the very center of all intellectual activities and formed the basis of philosophical training. The gate of Plato's Academy in Athens is said to have borne the inscription, "You are not allowed to enter here unless you know geometry." The Greeks believed that their mathematical theorems were expressions of eternal and exact truths about the real world, and that geometrical shapes were manifestations of absolute beauty. Geometry was considered to be the perfect combination of logic and beauty and was thus believed to be of divine origin. Hence Plato's dictum, "God is a geometer."

Since geometry was seen as the revelation of God, it was obvious to the Greeks that the heavens should exhibit perfect geometrical shapes. This meant that the heavenly bodies had to move in circles. To present the picture as being even more geometrical they were thought to be fixed to a series of concentric crystalline spheres which moved as a whole, with the earth at the center.

In subsequent centuries, Greek geometry continued to exert a strong influence on Western philosophy and science. Euclid's *Elements* was a standard textbook in European schools until the beginning of this century, and Euclidean geometry was taken to be the true nature of space for more than two thousand years. It took an Einstein to make scientists and philosophers realize that geometry is not inherent in nature, but is imposed upon it by the mind. In the words of Henry Margenau:

> The central recognition of the theory of relativity is that geometry . . . is a construct of the intellect. Only when this discovery is accepted can the mind feel free to tamper with the time-honored notions of space and time, to survey the range of possibilities available for defining them, and to select that formulation which agrees with observation.[1]

Eastern philosophy, unlike that of the Greeks, has always maintained that space and time are constructs of the mind. The Eastern mystics treated them like all other intellectual concepts; as relative, limited, and illusory. In a Buddhist text, for example, we find the words,

> It was taught by the Buddha, oh monks, that . . . the past, the future, physical space, . . . and individuals are nothing but names, forms of thought, words of common usage, merely superficial realities.[2]

Thus in the Far East, geometry never attained the status it had in ancient Greece, although this does not mean that the Indians and Chinese had little knowledge of it. They used it extensively in building altars of precise geometrical shapes, in measuring the land and mapping out the heavens, but never to determine abstract and eternal truths. This philosophical attitude is also reflected in the fact that ancient Eastern science generally did not find it necessary to fit nature into a scheme of straight lines and perfect circles. Joseph Needham's remarks about Chinese astronomy are very interesting in this connection:

> The Chinese [astronomers] did not feel the need for [geometrical] forms of explanation—the component organisms in the universal organism followed their Tao each according to its own nature, and their motions could be dealt with in the essentially "non-representational" form of algebra. The Chinese were thus free from that obsession of European astronomers for the circle as the most perfect figure, . . . nor did they experience the medieval prison of the crystalline spheres.[3]

Thus the ancient Eastern philosophers and scientists already had the attitude which is so basic to relativity theory—that our notions of geometry are not absolute and unchangeable properties of nature, but intellectual constructions. In the words of Ashvaghosha:

> Be it clearly understood that space is nothing but a mode of particularization and that it has no real existence of its own. . . . Space exists only in relation to our particularizing consciousness.[4]

The same applies to our idea of time. The Eastern mystics link the notions of both space and time to particular states of consciousness. Being able to go beyond the ordinary state through meditation, they have realized that the conventional notions of space and time are not the ultimate truth. The refined notions of space and time resulting from their mystical experiences appear to be in many ways similar to the notions of modern physics, as exemplified by the theory of relativity.

What, then, is this new view of space and time which emerged from relativity theory? It is based on the discovery that all space and time measurements are relative. The relativity of spatial specifications was, of course, nothing new. It was well known before Einstein that the position of an object in space can be defined only relative to some other object. This is usually done with the help of three coordinates and the point from which the coordinates are measured may be called the location of the "observer."

To illustrate the relativity of such coordinates, imagine two observers floating in space and observing an umbrella, as drawn opposite. Observer A sees the umbrella to his left and slightly inclined, so that the upper end is nearer to him. Observer B, on the other hand, sees the umbrella to his right and in such a way that the upper end is farther away. By extending this two-dimensional example to three dimensions, it becomes clear that all spatial specifications—such as "left," "right," "up," "down," "oblique," etc.—depend on the position of the observer and are thus relative. This was known long before relativity theory. As far as time is concerned, however, the

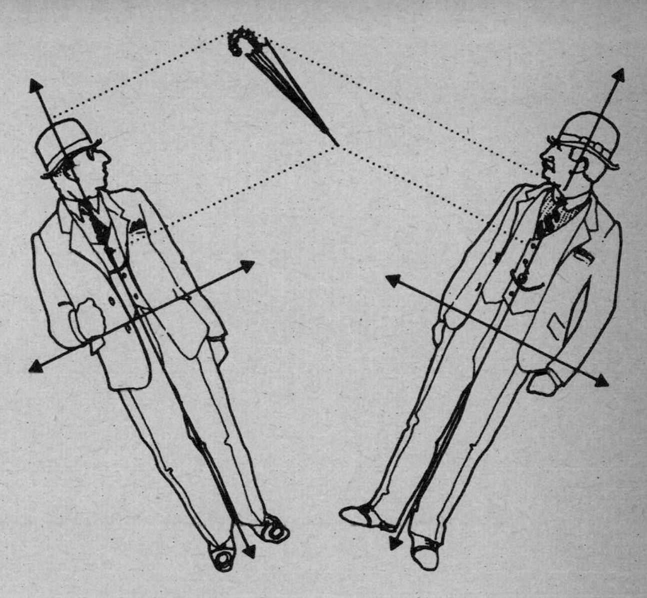

Two observers, A and B, observing an umbrella

situation in classical physics was entirely different. The temporal order of two events was assumed to be independent of any observer. Specifications referring to time —such as "before," "after," or "simultaneous"—were thought to have an absolute meaning independent of any coordinate system.

Einstein recognized that temporal specifications, too, are relative and depend on the observer. In everyday life, the impression that we can arrange the events around us in a unique time sequence is created by the fact that the velocity of light—186,000 miles per second—is so high, compared to any other velocity we experience, that we can assume we are observing events at the instant they are occurring. This, however, is incorrect. Light needs some time to travel from the event to the observer. Normally, this time is so short that the propagation of light can be considered to be instantaneous; but when the observer moves with a high velocity with respect to the observed phenomena, the time span between the occurrence of an event and its observation plays a crucial role in establishing a sequence of events. Einstein realized that in such a case, observers moving at different velocities will order events differently in time.* Two events which are seen as occurring simultaneously by one observer may occur in different temporal sequences for others. For ordinary velocities, the differences are so small that they cannot be detected, but when the velocities approach the speed of light, they give rise to measurable effects. In high energy physics, where the events are interactions between particles moving almost at the speed of light, the relativity of time is well established and has been confirmed by countless experiments.**

The relativity of time also forces us to abandon the Newtonian concept of an absolute space. Such a space was seen as containing a definite configuration of matter at every instant; but now that simultaneity is seen to be a relative concept, depending on the state of motion of the observer, it is no longer possible to define such a definite instant for the whole universe. A distant event which takes place at some particular instant for one observer may happen earlier or later for another observer. It is therefore not possible to speak about "the universe at a given instant" in an absolute way; there is no absolute space independent of the observer.

Relativity theory has thus shown that all measurements involving space and time lose their absolute significance and has forced us to abandon the classical concepts of an absolute space and an absolute time. The fundamental importance of this development has been clearly expressed by Mendel Sachs in the following words:

> The real revolution that came with Einstein's theory . . . was the abandonment of the idea that the

* To derive this result it is essential to take into account the fact that the speed of light is the same for all observers.

** Note that in this case, the observer is at rest in his laboratory, but the events he observes are caused by particles moving at different velocities. The effect is the same. What counts is the *relative* motion of observer and observed events. Which of the two moves with respect to the laboratory is irrelevant.

> space-time coordinate system has objective significance as a separate physical entity. Instead of this idea, relativity theory implies that the space and time coordinates are only the elements of a language that is used by an observer to describe his environment.[5]

This statement from a contemporary physicist shows the close affinity between the notions of space and time in modern physics and those held by the Eastern mystics who say, as quoted before, that space and time "are nothing but names, forms of thought, words of common usage."

Since space and time are now reduced to the subjective role of the elements of the language a particular observer uses for his or her description of natural phenomena, each observer will describe the phenomena in a different way. To abstract some universal natural laws from their descriptions, they have to formulate these laws in such a way that they have the same form in all coordinate systems, i.e., for all observers in arbitrary positions and relative motion. This requirement is known as the principle of relativity and was, in fact, the starting point of relativity theory. It is interesting that the germ of the theory of relativity was contained in a paradox which occurred to Einstein when he was only sixteen. He tried to imagine how a beam of light would look to an observer who traveled along with it at the speed of light, and he concluded that such an observer would see the beam of light as an electromagnetic field oscillating back and forth without moving on, i.e., without forming a wave. Such a phenomenon, however, is unknown in physics. It seemed thus to the young Einstein that something which was observed by one observer to be a well-known electromagnetic phenomenon, namely a light wave, would appear as a phenomenon contradicting the laws of physics to another observer, and this he could not accept. In later years, Einstein realized that the principle of relativity can be satisfied in the description of electromagnetic phenomena only if all spatial *and* temporal specifications are relative. The laws of mechanics, which govern the phenomena associated with moving bodies, and the laws of electrodynamics, the theory of electricity and magnetism, can then be formulated in a common "relativistic" framework which incorporates time with the three space coordinates as a fourth coordinate to be specified relative to the observer.

In order to check whether the principle of relativity is satisfied, that is, whether the equations of one's theory look the same in all coordinate systems, one must of course be able to translate the space and time specifications from one coordinate system, or "frame of reference," to the other. Such translations, or "transformations" as they are called, were already well known and widely used in classical physics. The transformation between the two reference frames pictured on p. 151, for example, expresses each of the two coordinates of observer A (one horizontal and one vertical, as indicated by the arrow-headed cross in the drawing) as a combination of the coordinates of observer B, and vice versa. The exact expressions can be easily obtained with the help of elementary geometry.

In relativistic physics, a new situation arises because time is added to the three space coordinates as a fourth dimension. Since the transformations between different frames of reference express each coordinate of one frame as a combination of the coordinates of the other frame, a space coordinate in one frame will in general appear as a mixture of space and time coordinates in another frame. This is indeed an entirely new situation. Every change of coordinate systems mixes space and time in a mathematically well-defined way. The two can therefore no longer be separated, because what is space to one observer will be a mixture of space and time to another. Relativity theory has shown that space is not three-dimensional and time is not a separate entity. Both are intimately and inseparably connected and form a four-dimensional continuum which is called "space-time." This concept of space-time was introduced by Hermann Minkowski in a famous lecture in 1908 with the following words:

> The views of space and time which I wish to lay before you have sprung from the soil of experimental physics, and therein lies their strength. They

> are radical. Henceforth space by itself, and time by itself, are doomed to fade away into mere shadows, and only a kind of union of the two will preserve an independent reality.[6]

The concepts of space and time are so basic for the description of natural phenomena that their modification entails an alteration of the whole framework we use in physics to describe nature. In the new framework, space and time are treated on an equal footing and are connected inseparably. In relativistic physics, we can never talk about space without talking about time, and vice versa. This new framework has to be used whenever phenomena involving high velocities are described.

The intimate link between space and time was well known in astronomy, in a different context, long before relativity theory. Astronomers and astrophysicists deal with extremely large distances, and here again the fact that light needs some time to travel from the observed object to the observer is important. Because of the finite velocity of light, the astronomer never looks at the universe in its present state, but always looks back into the past. It takes light eight minutes to travel from the sun to the earth, and hence we see the sun, at any moment, as it existed eight minutes ago. Similarly, we see the nearest star as it existed four years ago, and with our powerful telescopes we can see galaxies as they existed millions of years ago.

The finite velocity of light is by no means a handicap for astronomers but is a great advantage. It allows them to observe the evolution of stars, star clusters, or galaxies at all stages just by looking out into space and back into time. All types of phenomena that happened during the past millions of years can actually be observed somewhere in the skies. Astronomers are thus used to the importance of the link between space and time. What relativity theory tells us is that this link is important not only when we deal with large distances, but also when we deal with high velocities. Even here on earth, the measurement of any distance is not independent of time, because it involves the specification of the observer's state of motion and thus a reference to time.

The unification of space and time entails—as mentioned in the previous chapter—a unification of other basic concepts, and this unifying aspect is the most characteristic feature of the relativistic framework. Concepts which seemed totally unrelated in nonrelativistic physics are now seen to be but different aspects of one and the same concept. This feature gives the relativistic framework great mathematical elegance and beauty. Many years of work with relativity theory have made us appreciate this elegance and become thoroughly familiar with the mathematical formalism. However, this has not helped our intuition very much. We have no direct sensory experience of the four-dimensional space-time, nor of the other relativistic concepts. Whenever we study natural phenomena involving high velocities, we find it very hard to deal with these concepts both at the level of intuition and ordinary language.

For example, in classical physics it was always assumed that rods in motion and at rest have the same length. Relativity theory has shown that this is not true. The length of an object depends on its motion relative to the observer and it changes with the velocity of that motion. The change is such that the object contracts in the direction of its motion. A rod has its maximum length in a frame of reference where it is at rest, and it becomes shorter with increasing velocity relative to the observer. In the "scattering" experiments of high-energy physics, where particles collide with extremely high velocities, the relativistic contraction is so extreme that spherical particles are reduced to "pancake" shapes.

It is important to realize that it makes no sense to ask which is the "real" length of an object, just as it makes no sense in our everyday life to ask for the real length of somebody's shadow. The shadow is a projection of points in three-dimensional space onto a two-dimensional plane, and its length will be different for different angles of projection. Similarly, the length of a moving object is the projection of points in four-dimensional space-time onto three-dimensional space, and its length is different in different frames of reference.

What is true for lengths is also true for time intervals. They, too, depend on the frame of reference, but contrary to spatial distances they become longer as the velocity relative to the observer increases. This means that clocks in motion run slower; time slows down. These clocks can be of varying types: mechanical clocks, atomic clocks, or even a human heartbeat. If one of two twins went on a fast round-trip into outer space, he would be younger than his brother when he came back home, because all his "clocks"—his heartbeat, bloodflow, brainwaves, etc.—would slow down during the journey, from the point of view of the man on the ground. The traveler himself, of course, would not notice anything unusual, but on his return he would suddenly realize that his twin brother was now much older. This "twin paradox" is perhaps the most famous paradox of modern physics. It has provoked heated discussions in scientific journals, some of which are still going on; an eloquent proof of the fact that the reality described by relativity theory cannot easily be grasped by our ordinary understanding.

The slowing down of clocks in motion, unbelievable as it sounds, is well tested in particle physics. Most of the subatomic particles are unstable, i.e., they disintegrate after a certain time into other particles. Numerous experiments have confirmed the fact that the lifetime* of such an unstable particle depends on its state of motion. It increases with the speed of the particle. Particles moving with 80 percent of the speed of light live about 1.7 times as long as their slow "twin brothers," and at 99 percent of the speed of light they live about 7 times as long. This, again, does not mean that the intrinsic lifetime of the particle changes. From the particle's point of view, its lifetime is always the same, but from the point of view of the laboratory observer the particle's "internal clock" has slowed down, and therefore it lives longer.

All these relativistic effects seem strange only because we cannot exprience the four-dimensional space-time world with our senses, but can only observe its three-dimensional "images." These images have different aspects in different frames of reference; moving objects

* A small technical point should perhaps be mentioned. When we speak about the lifetime of a certain kind of unstable particle, we always mean the average lifetime. Due to the statistical character of subatomic physics, we cannot make any statement about individual particles.

look different from objects at rest, and moving clocks run at a different rate. These effects will seem paradoxical if we do not realize that they are only the projections of four-dimensional phenomena, just as shadows are projections of three-dimensional objects. If we could visualize the four-dimensional space-time reality, there would be nothing paradoxical at all.

The Eastern mystics, as mentioned above, seem to be able to attain nonordinary states of consciousness in which they transcend the three-dimensional world of everyday life to experience a higher, multidimensional reality. Thus Aurobindo speaks about "a subtle change which makes the sight see in a sort of fourth dimension."[7] The dimensions of these states of consciousness may not be the same as the ones we are dealing with in relativistic physics, but it is striking that they have led the mystics toward notions of space and time which are very similar to those implied by relativity theory.

Throughout Eastern mysticism, there seems to be a strong intuition for the "space-time" character of reality. The fact that space and time are inseparably linked, which is so characteristic of relativistic physics, is stressed again and again. This intuitive notion of space and time has, perhaps, found its clearest expression and its most far-reaching elaboration in Buddhism, and in particular in the *Avatamsaka* school of Mahayana Buddhism. The *Avatamsaka Sutra*, on which this school is based,* gives a vivid description of how the world is experienced in the state of enlightenment. The awareness of an 'interpenetration of space and time'—a perfect expression to describe space-time—is repeatedly emphasized in the *sutra* and is seen as an essential characteristic of the enlightened state of mind. In the words of D. T. Suzuki:

> The significance of the *Avatamsaka* and its philosophy is unintelligible unless we once experience . . . a state of complete dissolution where there is no more distinction between mind and body, subject and object. . . . We look around and perceive that . . . every object is related to every other object . . .

* See p. 89.

> not only spatially, but temporally. . . . As a fact of pure experience, there is no space without time, no time without space; they are interpenetrating.[8]

One could hardly find a better way of describing the relativistic concept of space-time. In comparing Suzuki's statement to the one quoted before by Minkowski, it is also interesting to note that both the physicist and the Buddhist emphasize the fact that their notions of space-time are based on experience; on scientific experiments in one case, and on mystical experience in the other.

In my opinion, the time-minded intuition of Eastern mysticism is one of the main reasons why its views of nature seem to correspond, in general, much better to modern scientific views than do those of most Greek philosophers. Greek natural philosophy was, on the whole, essentially static and largely based on geometrical considerations. It was, one could say, extremely 'non-relativistic,' and its strong influence on Western thought may well be one of the reasons why we have such great conceptual difficulties with relativistic models in modern physics. The Eastern philosophies, on the other hand, are 'space-time' philosophies, and thus their intuition often comes very close to the views of nature implied by our modern relativistic theories.

Because of the awareness that space and time are intimately connected and interpenetrating, the world views of modern physics and of Eastern mysticism are both intrinsically dynamic views which contain time and change as essential elements. This point will be discussed in detail in the following chapter, and constitutes the second main theme recurring throughout this comparison of physics and Eastern mysticism, the first being the unity of all things and events. As we study the relativistic models and theories of modern physics, we shall see that all of them are impressive illustrations of the two basic elements of the Eastern world-view—the basic oneness of the universe and its intrinsically dynamic character.

The theory of relativity discussed so far is known as the "special theory of relativity." It provides a common framework for the description of the phenomena associated with moving bodies and with electricity and magnetism, the basic features of this framework being the relativity of space and time and their unification into four-dimensional space-time.

In the "general theory of relativity," the framework of the special theory is extended to include gravity. The effect of gravity, according to general relativity, is to make space-time curved. This, again, is extremely hard to imagine. We can easily imagine a two-dimensional curved surface, such as the surface of an egg, because we can see such curved surfaces lying in three-dimensional space. The meaning of the word "curvature" for two-dimensional curved surfaces is thus quite clear; but when it comes to three-dimensional space—let alone four-dimensional space-time—our imagination abandons us. Since we cannot look at three-dimensional space "from outside," we cannot imagine how it can be "bent in some direction."

To understand the meaning of curved space-time, we have to use curved two-dimensional surfaces as analogies. Imagine, for example, the surface of a sphere. The crucial fact which makes the analogy to space-time possible is that the curvature is an intrinsic property of that surface and can be measured without going into three-dimensional space. A two-dimensional insect confined to the surface of the sphere and unable to experience three-dimensional space could nevertheless find out that the surface on which he is living is curved, provided that he can make geometrical measurements.

To see how this works, we have to compare the geometry of our bug on the sphere with that of a

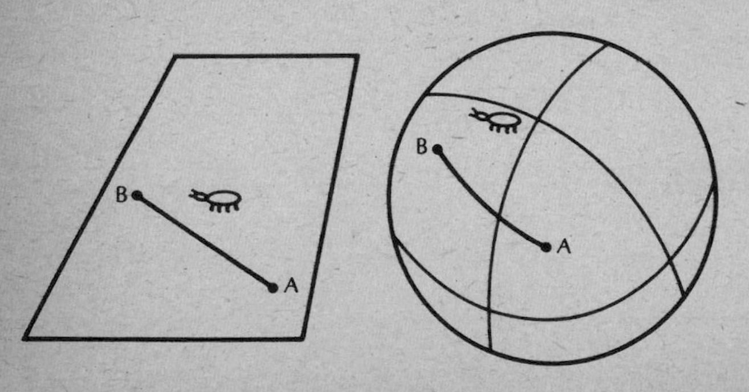

drawing a "straight line" on a plane and on a sphere

similar insect living on a flat surface.* Suppose the two bugs begin their study of geometry by drawing a straight line, defined as the shortest connection between two points. The result is shown opposite. We see that the bug on the flat surface drew a very nice straight line; but what did the bug on the sphere do? For him, the line he drew is the shortest connection between the two points A and B, since any other line he may draw will be longer; but from our point of view we recognize it as a curve (the arc of a great circle, to be precise). Now suppose that the two bugs study triangles. The bug on the plane will find that the three angles of any triangle add up to two right angles, i.e., to 180°; but the bug on the sphere will discover that the sum of the angles in his triangles is

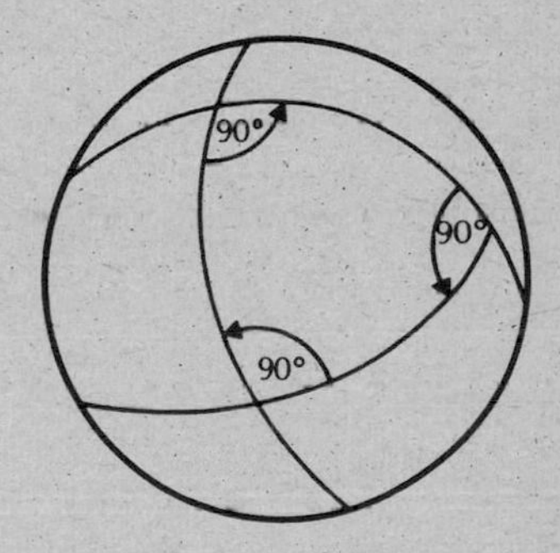

on a sphere a triangle can have three right angles

always greater than 180°. For small triangles, the excess is small, but it increases as the triangles become larger; and as an extreme case, our bug on the sphere will even be able to draw triangles with three right angles. Finally, let the two bugs draw circles and measure their circumference. The bug on the plane will find that the circumference is always equal to 2π times the radius, independent of the size of the circle. The bug on the sphere, on the other hand, will notice that the circumference is always less than 2π times the radius. As can

* The following examples are taken from R. P. Feynman, R. B. Leighton and M. Sands, *The Feynman Lectures on Physics* (Addison-Wesley, Reading, Mass., 1966), Vol. II, ch. 42.

be seen in the figure below, our three-dimensional point of view allows us to see that what the bug calls the radius of his circle is in fact a curve which is always longer than the true radius of the circle.

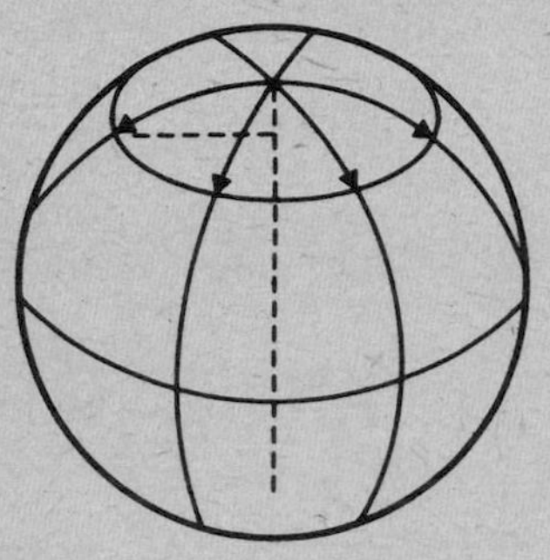

drawing a circle on a sphere

As the two insects continue to study geometry, the one on the plane should discover the axioms and laws of Euclidean geometry, but his colleague on the sphere will discover different laws. The difference will be small for small geometrical figures, but will increase as the figures become larger. The example of the two bugs shows that we can always determine whether a surface is curved or not, just by making geometrical measurements on the surface, and by comparing the results with those predicted by Euclidean geometry. If there is a discrepancy, the surface is curved; and the larger the discrepancy is—for a given size of figures—the stronger the curvature.

In the same way, we can define a curved three-dimensional space to be one in which Euclidean geometry is no longer valid. The laws of geometry in such a space will be of a different, "non-Euclidean" type. Such a non-Euclidean geometry was introduced as a purely abstract mathematical idea in the nineteenth century by the mathematician Georg Riemann, and it was not considered to be more than that until Einstein made the revolutionary suggestion that the three-dimensional space in which we live is actually curved. According to Einstein's theory, the curvature of space is caused by the gravitational fields of massive bodies. Wherever there is a massive object, the space around it is curved, and the degree of curvature, that is, the degree to which the geometry deviates from that of Euclid, depends on the mass of the object.

The equations relating the curvature of space to the distribution of matter in that space are called Einstein's field equations. They can be applied not only to determine the local variations of curvature in the neighborhood of stars and planets, but also to find out whether there is an overall curvature of space on a large scale. In other words, Einstein's equations can be used to determine the structure of the universe as a whole. Unfortunately, they do not give a unique answer. Several mathematical solutions of the equations are possible, and these solutions constitute the various models of the universe studied in cosmology, some of which will be discussed in the following chapter. To determine which of them corresponds to the actual structure of our universe is the main task of present-day cosmology.

Since space can never be separated from time in relativity theory, the curvature caused by gravity cannot be limited to three-dimensional space, but must extend to four-dimensional space-time, and this is, indeed, what the general theory of relativity predicts. In a curved space-time, the distortions caused by the curvature affect not only the spatial relationships described by geometry but also the lengths of time intervals. Times does not flow at the same rate as in "flat space-time," and as the curvature varies from place to place, according to the distribution of massive bodies, so does the flow of time. It is important to realize, however, that this variation of the flow of time can only be seen by an observer who remains in a different place from the clocks used to measure the variation. If the observer, for example, went to a place where time flows more slowly, all her clocks would slow down, too, and she would have no means of measuring the effect.

In our terrestrial environment, the effects of gravity on space and time are so small that they are insignificant, but in astrophysics, which deals with extremely massive bodies, like planets, stars, and galaxies, the curvature of space-time is an important phenomenon. All observations have so far confirmed Einstein's theory and thus force us to believe that space-time is indeed curved. The most extreme effects of the curvature of space-time become apparent during the gravitational collapse of a massive star. According to current ideas in astrophysics, every star reaches a stage in its evolution where it collapses due to the mutual gravitational attraction of its particles. Since this attraction increases rapidly as the distance between the particles decreases, the collapse accelerates; and if the star is massive enough—that is, if it is more than twice as massive as the sun—no known process can prevent the collapse from going on indefinitely.

As the star collapses and becomes more and more dense, the force of gravity on its surface becomes stronger and stronger, and consequently the space-time around it becomes more and more curved. Because of the increasing force of gravity on the star's surface, it becomes more and more difficult to get away from it, and eventually the star reaches a stage where nothing—not even light—can escape from its surface. At that stage, we say that an "event horizon" forms around the star because no signal can get away from it to communicate any event to the outside world. The space around the star is then so strongly curved that all the light is trapped in it and cannot escape. We are not able to see such a star because its light can never reach us. For this reason it is called a black hole. The existence of black holes was predicted on the grounds of relativity theory as early as 1916, and black holes have lately received a great deal of attention because some recently discovered stellar phenomena might indicate the existence of a heavy star moving around some unseen partner which could be a black hole.

Black holes are among the most mysterious and most fascinating objects investigated by modern astrophysics and illustrate the effects of relativity theory in a most spectacular way. The strong curvature of space-time around them prevents not only all their light from reaching us, but has an equally striking effect on time. If a clock, flashing its signals to us were attached to the surface of the collapsing star, we would observe these signals to slow down as the star approached the event horizon, and once the star had become a black hole, no clock signals would reach us anymore. To an outside observer, the flow of time on the star's surface slows down as the star collapses and stops altogether at the event horizon. Therefore, the complete collapse of the star takes an infinite time. The star itself, however, experiences nothing peculiar when it collapses beyond the event horizon. Time continues to flow normally, and the collapse is completed after a finite period of time, when the star has contracted to a point of infinite density. So how long does the collapse really take: a finite time or an infinite time? In the world of relativity theory, such a question does not make sense. The lifetime of a collapsing star, like all other time spans, is relative and depends on the frame of reference of the observer.

In the general theory of relativity, the classical concepts of space and time as absolute and independent entities are completely abolished. Not only are all measurements involving space and time relative, depending on the state of motion of the observer, but the whole structure of space-time is inextricably linked to the distribution of matter. Space is curved to different degrees, and time flows at different rates in different parts of the universe. We have thus come to apprehend that our notions of a three-dimensional Euclidean space and of linear flowing time are limited to our ordinary experience of the physical world and have to be completely abandoned when we extend this experience.

The Eastern sages, too, talk about an extension of their experience of the world in higher states of consciousness, and they affirm that these states involve a radically different experience of space and time. They emphasize not only that they go beyond ordinary three-dimensional space in meditation, but also—and even more forcefully—that the ordinary awareness of time is transcended. Instead of a linear succession of instants, they experience—so they say—an infinite, timeless, and yet dynamic present. In the following passages, three Eastern mystics speak about the experience of this 'eternal now'; Chuang Tzu, the Taoist sage; Hui-neng, the Sixth Zen Patriarch; and D. T. Suzuki, the contemporary Buddhist scholar.

> Let us forget the lapse of time; let us forget the conflict of opinions. Let us make our appeal to the infinite, and take up our positions there.[9]
>
> *Chuang Tzu*

> The absolute tranquillity is the present moment. Though it is at this moment, there is no limit to this moment, and herein is eternal delight.[10]
>
> *Hui-neng*

> In this spiritual world there are no time divisions such as the past, present and future; for they have contracted themselves into a single moment of the present where life quivers in its true sense. . . . The past and the future are both rolled up in this present moment of illumination, and this present moment is not something standing still with all its contents, for it ceaselessly moves on.[11]
>
> *D. T. Suzuki*

Talking about an experience of timeless present is almost impossible, because all words like "timeless," "present," "past," "moment," etc., refer to the conventional notions of time. It is thus extremely difficult to understand what the mystics mean in passages like those quoted; but here again, modern physics may facilitate the understanding, as it can be used to illustrate graphically how its theories transcend ordinary notions of time.

In relativistic physics, the history of an object, say a particle, can be represented in a so-called "space-time diagram" (see figure below). In these diagrams, the horizontal direction represents space,* and the vertical

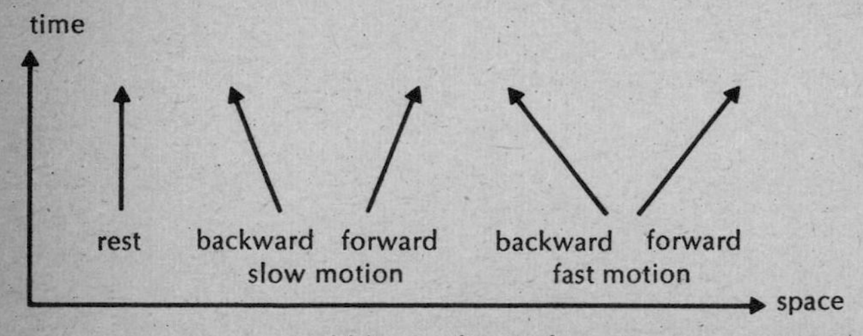

world lines of particles

* Space, in these diagrams, has only one dimension; the other two dimensions have to be suppressed to make a plane diagram possible.

direction, time. The path of the particle through space-time is called its "world line." If the particle is at rest, it nevertheless moves through time, and its world line is, in that case, a straight vertical line. If the particle moves in space, its world line will be inclined; the greater the inclination of the world line, the faster the particle moves. Note that the particles can move only upward in time, but can move backward or forward in space. Their world lines can be inclined toward the horizontal to various degrees, but can never become completely horizontal, since this would mean that a particle travels from one place to the other in no time at all.

Space-time diagrams are used in relativistic physics to picture the interactions between various particles. For each process, we can draw a diagram and associate a definite mathematical expression with it which gives us the probability for that process to occur. The collision, or "scattering," process between an electron and a photon, for example, may be represented by a diagram like the one below. This diagram is read in the following way (from the bottom to the top, according to the direction of time): an electron (denoted by e^- because of its negative charge) collides with a photon (denoted by γ—"gamma"); the photon is absorbed by the electron which continues

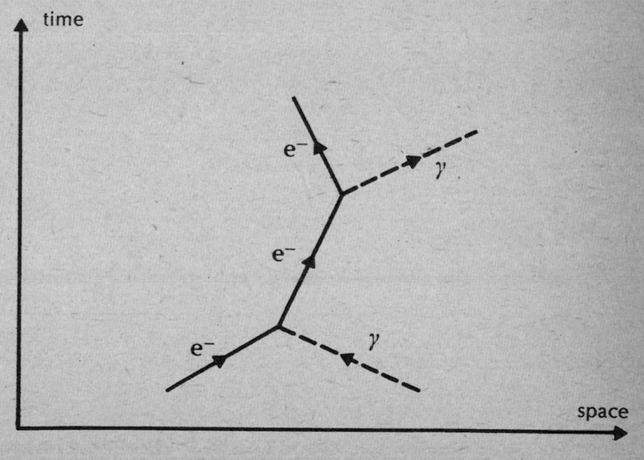

electron-photon scattering

its path with a different velocity (different inclination of the world line); after a while, the electron emits the photon again and reverses its direction of motion.

The theory which constitutes the proper framework for these space-time diagrams, and for the mathematical expressions associated with them, is called "quantum field theory." It is one of the main relativistic theories of modern physics whose basic concepts will be discussed later on. For our discussion of space-time diagrams, it will be sufficient to know two characteristic features of the theory. The first is the fact that all interactions involve the creation and destruction of particles, like the absorption and emission of the photon in our diagram; and the second feature is a basic symmetry between particles and antiparticles. For every particle, there exists an antiparticle with equal mass and opposite charge. The antiparticle of the electron, for example, is called the positron and is usually denoted by e^+. The photon, having no charge, is its own antiparticle. Pairs of electrons and positrons can be created spontaneously by photons, and can be made to turn into photons in the reverse process of annihilation.

The space-time diagrams, now, are greatly simplified if the following trick is adopted. The arrowhead on a world line is no longer used to indicate the direction of motion of the particle (which is unnecessary, anyway, since all particles move forward in time, i.e., upward in the diagram). Instead, the arrowhead is used to distinguish between particles and antiparticles: if it points upward, it indicates a particle (e.g. an electron), if it points downward, an antiparticle (e.g. a positron). The photon, being its own antiparticle, is represented by a world line without any arrowhead. With this modification, we can now omit all the labels in our diagram without causing any confusion: the lines with arrowheads represent electrons, those without arrowheads, photons. To make the diagram even simpler, we can also leave out the space axis and the time axis, remembering that the direction of time is from the bottom to the top, and that the forward direction in space is from left to right. The resulting space-time diagram for the electron-photon scattering process looks then as follows.

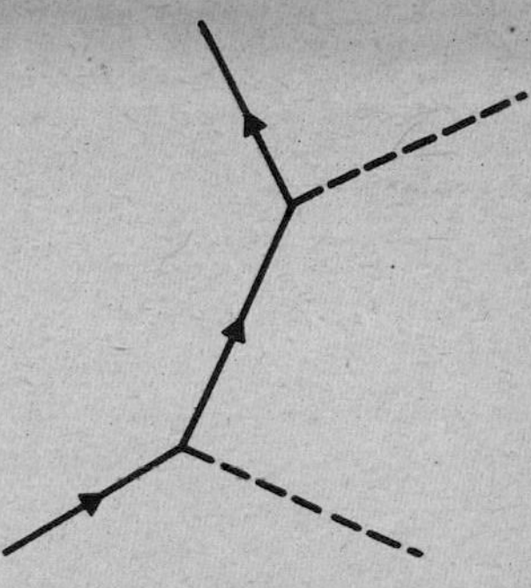

electron-photon scattering

If we want to picture the scattering process between a photon and a positron, we can draw the same diagram and just reverse the direction of the arrowheads:

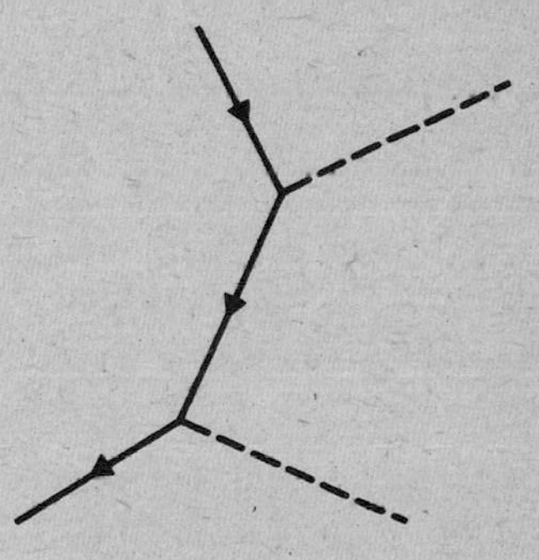

positron-photon scattering

So far, there has been nothing unusual in our discussion of space-time diagrams. We have read them from the bottom to the top, according to our conventional notion of a linear flow of time. The unusual aspect is connected with diagrams containing positron lines, like the one picturing the positron-photon scattering. The mathematical formalism of field theory suggests that these lines can be interpreted in two ways; either as positrons moving forward in time, or as electrons *moving backward in time*! The interpretations are mathematically identical;

the same expression describes an antiparticle moving from the past to the future, or a particle moving from the future to the past. Our two diagrams can thus be

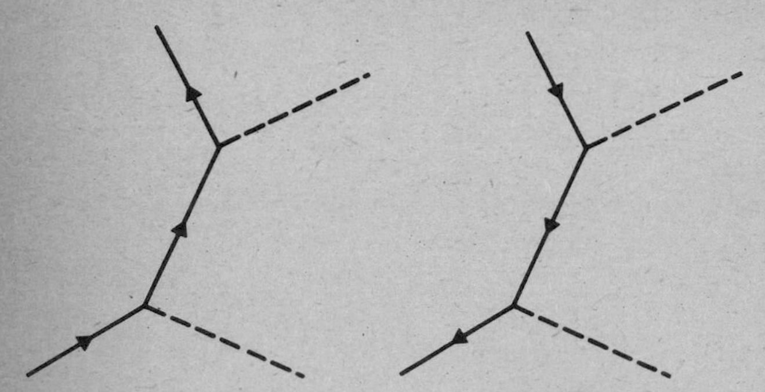

seen as picturing the same process evolving in different directions in time. Both of them can be interpreted as the scattering of electrons and photons, but in one process the particles move forward in time, in the other they move backward.* The relativistic theory of particle interactions shows thus a complete symmetry with regard to the direction of time. All space-time diagrams may be read in either direction. For every process, there is an equivalent process with the direction of time reversed and particles replaced by antiparticles.**

To see how this surprising feature of the world of subatomic particles affects our views of space and time, consider the process shown in the diagram opposite. Reading the diagram in the conventional way, from the bottom to the top, we will interpret it as follows: an electron (represented by a solid line) and a photon (represented by a dashed line) approach each other; the photon creates an electron-positron pair at point A, the electron flying off to the right, the positron to the left;

* The dashed lines are always interpreted as photons, whether they move forward or backward in time, because the antiparticle of a photon is again a photon.

** Recent experimental evidence suggests that this might not be true for a particular process involving a "super-weak interaction." Apart from this process, for which the role of time-reversal symmetry is not yet clear, all particle interactions seem to show a basic symmetry with regard to the direction of time.

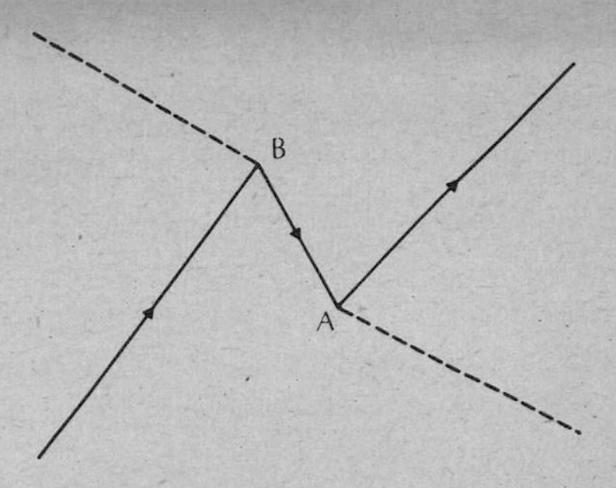

scattering process involving photons,
electrons and a positron

the positron then collides with the initial electron at point B and they annihilate each other, creating a photon in the process which flies off to the left. Alternatively, we may also interpret the process as the interaction of the two photons with a single electron traveling first forward in time, then backward, and then forward again. For this interpretation, we just follow the arrows on the electron line all the way through; the electron travels to point B where it emits a photon and reverses its direction to travel backward through time to point A; there it absorbs the initial photon, reverses its direction again and flies off traveling forward through time. In a way, the second interpretation is much simpler because we just follow the world line of one particle. On the other hand, we notice immediately that in doing so we run into serious difficulties of language. The electron travels "first" to point B, and "then" to A; yet the absorption of the photon at A happens before the emission of the other photon at B.

The best way to avoid these difficulties is to see space-time diagrams like the one above not as chronological records of the paths of particles through time, but rather as four-dimensional patterns in space-time representing a network of interrelated events which does not have any definite direction of time attached to it. Since all particles can move forward and backward in time, just as they can move left and right in space, it does not make sense to impose a one-way flow of time on the diagrams. They are simply four-dimensional maps traced out in space-time in such a way that we cannot speak of any temporal sequence. In the words of Louis de Broglie:

> In space-time, everything which for each of us constitutes the past, the present, and the future is given *en bloc*. . . . Each observer, as his time passes, discovers, so to speak, new slices of space-time which appear to him as successive aspects of the material world, though in reality the ensemble of events constituting space-time exist prior to his knowledge of them.[12]

This, then, is the full meaning of space-time in relativistic physics. Space and time are fully equivalent; they are unified into a four-dimensional continuum in which the particle interactions can stretch in any direction. If we want to picture these interactions, we have to picture them in one "four-dimensional snapshot" covering the whole span of time as well as the whole region of space. To get the right feeling for the relativistic world of particles, we must "forget the lapse of time," as Chuang Tzu says, and this is why the space-time diagrams of field theory can be a useful analogy to the space-time experience of the Eastern mystic. The relevance of the analogy is made evident by the following remarks by Lama Govinda concerning Buddhist meditation:

> If we speak of the space-experience in meditation, we are dealing with an entirely different dimension. . . . In this space-experience the temporal sequence is converted into a simultaneous coexistence, the side-by-side existence of things . . . and this again does not remain static but becomes a living continuum in which time and space are integrated.[13]

Although the physicists use their mathematical formalism and their diagrams to picture interactions *en bloc* in four-dimensional space-time, they say that in the actual world each observer can experience the phenomena only in a succession of space-time sections: that is, in a temporal sequence. The mystics, on the other hand, maintain that they can actually experience the full span of space-time where time does not flow any longer. Thus the Zen master Dogen:

> It is believed by most that time passes; in actual fact, it stays where it is. This idea of passing may be called time, but it is an incorrect idea, for since one sees it only as passing, one cannot understand that it stays just where it is.[14]

Many of the Eastern teachers emphasize that thought must take place in time, but that vision can transcend it. "Vision," says Govinda, "is bound up with a space of a higher dimension, and therefore timeless."[15] The space-time of relativistic physics is a similar timeless space of a higher dimension. All events in it are interconnected, but the connections are not causal. Particle interactions can be interpreted in terms of cause and effect only when the space-time diagrams are read in a definite direction: e.g., from the bottom to the top. When they are taken as four-dimensional patterns without any definite direction of time attached to them, there is no "before" and no "after," and thus no causation.

Similarly, the Eastern mystics assert that in transcending time, they also transcend the world of cause and effect. Like our ordinary notions of space and time, causation is an idea which is limited to a certain experience of the world and has to be abandoned when this experience is extended. In the words of Swami Vivekananda:

> Time, space, and causation are like the glass through which the Absolute is seen. . . . In the Absolute there is neither time, space, nor causation.[16]

The Eastern spiritual traditions show their followers various ways of going beyond the ordinary experience of time and of freeing themselves from the chain of cause and effect—from the bondage of *karma*, as the Hindus and Buddhists say. It has therefore been said that Eastern mysticism is a liberation from time. In a way, the same may be said of relativistic physics.

Diagram of Change from the Taoist Canon, Northern Sung dynasty.

13 THE DYNAMIC UNIVERSE

The central aim of Eastern mysticism is to experience all phenomena in the world as manifestations of the same ultimate reality. This reality is seen as the essence of the universe, underlying and unifying the multitude of things and events we observe. The Hindus call it *Brahman*, the Buddhists *Dharmakaya* (the Body of Being), or *Tathata* (Suchness), and the Taoists *Tao*; each affirming that it transcends our intellectual concepts and defies further description. This ultimate essence, however, cannot be separated from its multiple manifestations. It is central to its very nature to manifest itself in myriad forms which come into being and disintegrate, transforming themselves into one another without end. In its phenomenal aspect, the cosmic One is thus intrinsically dynamic, and the apprehension of its dynamic nature is basic to all schools of Eastern mysticism. Thus D. T. Suzuki writes about the Kegon school of Mahayana Buddhism:

> The central idea of Kegon is to grasp the universe dynamically whose characteristic is always to move onward, to be forever in the mood of moving, which is life.[1]

This emphasis on movement, flow, and change is not only characteristic of the Eastern mystical traditions, but has been an essential aspect of the world view of mystics throughout the ages. In ancient Greece, Heraclitus taught that "everything flows" and compared the world to an ever-living fire, and in Mexico, the Yaqui mystic Don Juan talks about the "fleeting world" and affirms that "to be a man of knowledge one needs to be light and fluid."[2]

In Indian philosophy, the main terms used by Hindus and Buddhists have dynamic connotations. The word *Brahman* is derived from the Sanskrit root *brih*—to grow—and thus suggests a reality which is dynamic and alive. In the words of S. Radhakrishnan, "The word *Brahman* means growth and is suggestive of life, motion, and progress."[3] The *Upanishads* refer to *Brahman* as "this unformed, immortal, moving,"[4] thus associating it with motion even though it transcends all forms.

The *Rig Veda* uses another term to express the dynamic nature of the universe: the term *Rita*. This word comes from the root *ri*—to move; its original meaning in the *Rig Veda* being "the course of all things," "the order of nature." It plays an important role in the legends of the *Veda* and is connected with all the Vedic gods. The order of nature was conceived by the Vedic seers, not as a static divine law, but as a dynamic principle which is inherent in the universe. This idea is not unlike the Chinese conception of *Tao*—'The Way'—as the way in which the universe works; i.e., the order of nature. Like the Vedic seers, the Chinese sages saw the world in terms of flow and change, and thus gave the idea of a cosmic order an essentially dynamic connotation. Both concepts, *Rita* and *Tao*, were later brought down from their original cosmic level to the human level and were interpreted in a moral sense; *Rita* as the universal law which all gods and men must obey, and *Tao* as the right way of life.

The Vedic concept of *Rita* anticipates the idea of *karma* which was developed later to express the dynamic interplay of all things and events. The word *karma* means "action" and denotes the "active," or dynamic, interrelation of all phenomena. In the words of the *Bhagavad Gita*, "All actions take place in time by the interweaving of the forces of nature."[5] The Buddha took up the traditional concept of *karma* and gave it a new meaning by extending the idea of dynamic interconnections to the sphere of human situations. *Karma* thus came to signify the never-ending chain of cause and effect in human life which the Buddha had broken in attaining the state of enlightenment.

Hinduism has also found many ways of expressing the dynamic nature of the universe in mythical language. Thus Krishna says in the *Gita*, "If I did not engage in action, these worlds would perish,"[6] and Shiva, the Cosmic Dancer, is perhaps the most perfect personification of the dynamic universe. Through his dance, Shiva sustains the manifold phenomena in the world, unifying all things by immersing them in his rhythm and making them participate in the dance—a magnificent image of the dynamic unity of the universe.

The general picture emerging from Hinduism is one of an organic, growing, and rhythmically moving cosmos; of a universe in which everything is fluid and ever-changing, all static forms being *maya*, that is, existing only as illusory concepts. This last idea—the impermanence of all forms—is the starting point of Buddhism. The Buddha taught that "all compounded things are impermanent," and that all suffering in the world arises from our trying to cling to fixed forms—objects, people or ideas—instead of accepting the world as it moves and changes. The dynamic world view lies thus at the very root of Buddhism. In the words of S. Radhakrishnan:

> A wonderful philosophy of dynamism was formulated by Buddha 2,500 years ago. . . . Impressed with the transitoriness of objects, the ceaseless mutation and transformation of things, Buddha formulated a philosophy of change. He reduces substances, souls, monads, things to forces, movements, sequences, and processes, and adopts a dynamic conception of reality.[7]

Buddhists call this world of ceaseless change *samsara*, which means, literally, "incessantly in motion"; and they affirm that there is nothing in it which is worth clinging to. So for the Buddhists, an enlightened being is one who does not resist the flow of life, but keeps moving with it. When the *Ch'an* monk Yün-men was asked, "What is the *Tao*?" he answered simply, "Walk on!" Accordingly, Buddhists also call the Buddha the *Tathagata*, or "the one who comes and goes thus." In Chinese philosophy, the flowing and ever-changing reality is called the *Tao* and is seen as a cosmic process in which all things are involved. Like the Buddhists, the Taoists say that one should not resist the flow, but should adapt one's actions to it. This, again, is characteristic of the sage—the enlightened being. If the Buddha is one who "comes and goes thus," the Taoist sage is one who "flows," as Huai Nan Tzu says,* "in the current of the *Tao*."

The more one studies the religious and philosophical texts of the Hindus, Buddhists and Taoists, the more it becomes apparent that in all of them the world is conceived in terms of movement, flow and change. This dynamic quality of Eastern philosophy seems to be one of its most important features. The Eastern mystics see the universe as an inseparable web, whose interconnections are dynamic and not static. The cosmic web is alive; it moves, grows and changes continually. Modern physics, too, has come to conceive of the universe as such a web of relations and, like Eastern mysticism, has recognized that this web is intrinsically dynamic. The dynamic aspect of matter arises in quantum theory as a consequence of the wave-nature of subatomic particles, and is even more essential in relativity theory, as we shall see, where the unification of space and time implies that the being of matter cannot be separated from its activity. The properties of subatomic particles can therefore only be understood in a dynamic context; in terms of movement, interaction and transformation.

According to quantum theory, particles are also waves, and this implies that they behave in a very peculiar way. Whenever a subatomic particle is confined to a small region of space, it reacts to this confinement by moving around. The smaller the region of confinement, the faster will the particle "jiggle" around in it. This behavior is a typical "quantum effect," a feature of the subatomic world which has no macroscopic analogy. To see how it comes about, we have to remember that particles are represented, in quantum theory, by wave packets. As discussed previously,** the length of such a wave packet

* See p. 108.
** See p. 148.

represents the uncertainty in the location of the particle. The following wave pattern, for example, corresponds to

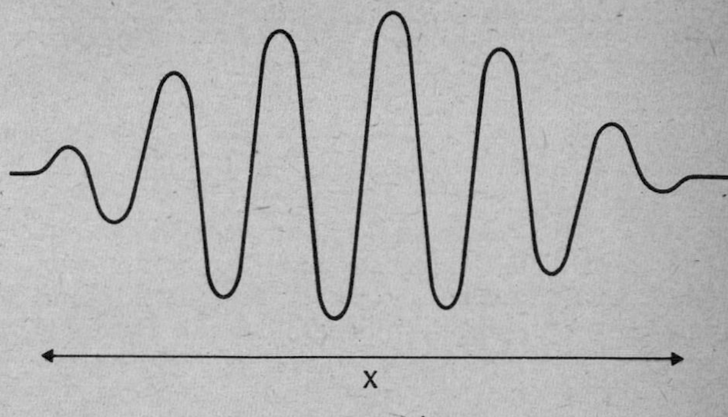

a wave packet

a particle located somewhere in the region X; where exactly we cannot say with certainty. If we want to localize the particle more precisely, i.e., if we want to confine it to a smaller region, we have to squeeze its wave packet into this region (see diagram below). This,

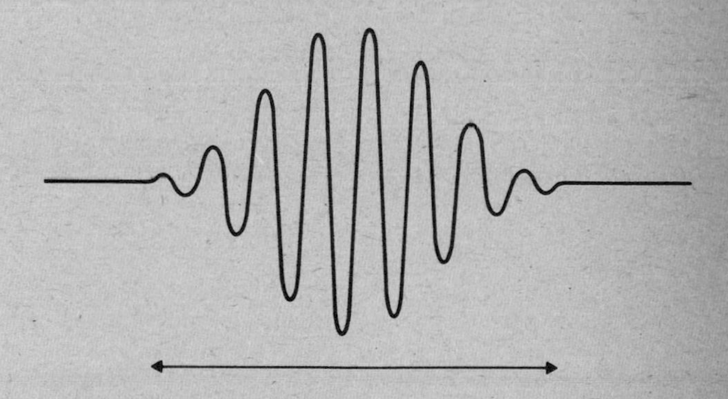

squeezing the wave packet into a smaller region

however, will affect the wavelength of the wave packet, and consequently the velocity of the particle. As a result, the particle will move around; the more it is confined, the faster it will move.

The tendency of particles to react to confinement with motion implies a fundamental "restlessness" of matter which is characteristic of the subatomic world. In this world, most of the material particles are bound to the molecular, atomic and nuclear structures, and therefore are not at rest but have an inherent tendency to move about—they are intrinsically restless. According to quantum theory, matter is thus never quiescent, but always in a state of motion. Macroscopically, the material objects around us may seem passive and inert, but when we magnify such a "dead" piece of stone or metal, we see that it is full of activity. The closer we look at it, the more alive it appears. All the material objects in our environment are made of atoms which link up with each other in various ways to form an enormous variety of molecular structures which are not rigid and motionless, but oscillate according to their temperature and in harmony with the thermal vibrations of their environment. In the vibrating atoms, the electrons are bound to the atomic nuclei by electric forces which try to keep them as close as possible, and they respond to this confinement by whirling around extremely fast. In the nuclei, finally, the protons and neutrons are pressed into a minute volume by the strong nuclear forces, and consequently race about with unimaginable velocities.

Modern physics, then, pictures matter not at all as passive and inert, but as being in a continuous dancing and vibrating motion whose rhythmic patterns are determined by the molecular, atomic and nuclear structures. This is also the way in which the Eastern mystics see the material world. They all emphasize that the universe has to be grasped dynamically, as it moves, vibrates and dances; that nature is not in a static, but a dynamic equilibrium. In the words of a Taoist text:

> The stillness in stillness is not the real stillness. Only when there is stillness in movement can the spiritual rhythm appear which pervades heaven and earth.[8]

In physics, we recognize the dynamic nature of the universe not only when we go to small dimensions—to the world of atoms and nuclei—but also when we turn to large dimensions—to the world of stars and galaxies.

Through our powerful telescopes we observe a universe in ceaseless motion. Rotating clouds of hydrogen gas contract to form stars, heating up in the process until they become burning fires in the sky. When they have reached that stage, they still continue to rotate, some of them ejecting material into space which spirals outwards and condenses into planets circling around the star. Eventually, after millions of years, when most of its hydrogen fuel is used up, a star expands, and then contracts again in the final gravitational collapse. This collapse may involve gigantic explosions, and may even turn the star into a black hole. All these activities—the formation of stars out of interstellar gas clouds, their contraction and subsequent expansion, and their final collapse—can actually be observed somewhere in the skies.

The spinning, contracting, expanding or exploding stars cluster into galaxies of various shapes—flat discs, spheres, spirals, etc.—which, again, are not motionless but rotate. Our galaxy, the Milky Way, is an immense disc of stars and gas turning in space like a huge wheel, so that all its stars—including the sun and its planets—move around the galaxy's center. The universe is, in fact, full of galaxies strewn through all the space we can see; all spinning like our own.

When we study the universe as a whole, with its millions of galaxies, we have reached the largest scale of space and time; and again, at that cosmic level, we discover that the universe is not static—it is expanding! This has been one of the most important discoveries in modern astronomy. A detailed analysis of the light received from distant galaxies has shown that the whole swarm of galaxies expands, and that it does so in a well-orchestrated way; the recession velocity of any galaxy we observe is proportional to the galaxy's distance. The more distant the galaxy, the faster it moves away from us; at double the distance, the recession velocity will also double. This is true not only for distances measured from our galaxy, but applies to any point of reference. Whichever galaxy you happen to be in, you will observe the other galaxies rushing away from you; nearby galaxies at several thousand miles per second, farther ones at higher speeds, and the farthest at velocities approaching the speed of light. The light from galaxies beyond that distance will never reach us, because they move away from us faster than the speed of light. Their light is—in the words of Sir Arthur Eddington—"like a runner on an expanding track with the winning post receding faster than he can run."

To have a better idea of the way in which the universe expands, we have to remember that the proper framework for studying its large-scale features is Einstein's general theory of relativity. According to this theory, space is not "flat," but is "curved," and the precise way in which it is curved is related to the distribution of matter by Einstein's field equations. These equations can be used to determine the structure of the universe as a whole; they are the starting point of modern cosmology.

When we talk about an expanding universe in the framework of general relativity, we mean an expansion in a higher dimension. Like the concept of curved space, we can only visualize such a concept with the help of a two-dimensional analogy. Imagine a balloon with a

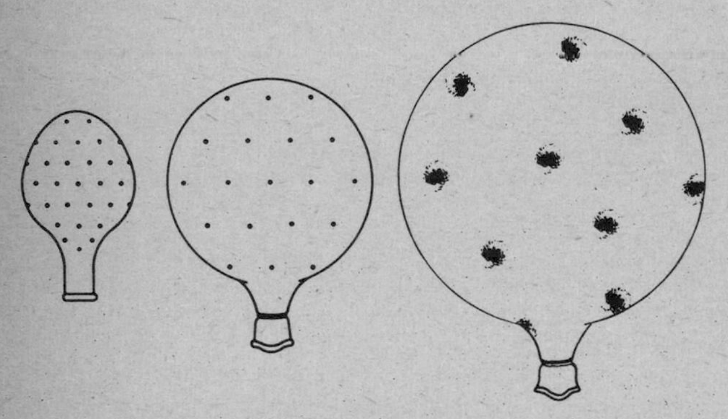

large number of dots on its surface. The balloon represents the universe, its two-dimensional curved surface representing the three-dimensional curved space, and the dots on the surface the galaxies in that space. When the balloon is blown up, all the distances between the dots increase. Whichever dot you choose to sit on, all the other dots will move away from you. The universe expands in the same way: whichever galaxy an observer happens to be in, the other galaxies will all move away from him.

An obvious question to be asked about the expanding universe is: how did it all start? From the relation between the distance of a galaxy and its recession velocity—which is known as Hubble's law—one can calculate the starting point of the expansion, in other words, the age of the universe. Assuming that there has been no change in the rate of expansion, which is by no means certain, one arrives at an age of the order of 10,000 million years. This, then, is the age of the universe. Most cosmologists believe today that the universe came into being in a highly dramatic event about 10,000 million years ago, when its total mass exploded out of a small primeval fireball. The present expansion of the universe is seen as the remaining thrust of this initial explosion. According to this "big-bang" model, the moment of the big bang marked the beginning of the universe and the beginning of space and time. If we want to know what happened before that moment, we run—again—into severe difficulties of thought and language. In the words of Sir Bernard Lovell:

> There we reach the great barrier of thought because we begin to struggle with the concepts of time and space before they existed in terms of our everyday experience. I feel as though I've suddenly driven into a great fog barrier where the familiar world has disappeared.[9]

As far as the future of the expanding universe is concerned, Einstein's equations do not provide a unique answer. They allow for several different solutions corresponding to different models of the universe. Some models predict that the expansion will continue forever; according to others, it is slowing down and will eventually change into a contraction. These models describe an oscillating universe, expanding for billions of years, then contracting until its total mass has condensed into a small ball of matter, then expanding again, and so on without end.

This idea of a periodically expanding and contracting universe, which involves a scale of time and space of vast proportions, has arisen not only in modern cosmology, but also in ancient Indian mythology. Experiencing the universe as an organic and rhythmically moving cosmos, the Hindus were able to develop evolutionary cosmologies which come very close to our modern scientific models. One of these cosmologies is based on the Hindu myth of *lila*—the divine play—in which *Brahman* transforms himself into the world.* *Lila* is a rhythmic play which goes on in endless cycles, the One becoming the many and the many returning into the One. In the *Bhagavad Gita*, the god Krishna describes this rhythmic play of creation in the following words:

> At the end of the night of time all things return to my nature; and when the new day of time begins I bring them again into light.
>
> Thus through my nature I bring forth all creation and this rolls around in the circles of time.
>
> But I am not bound by this vast work of creation. I am and I watch the drama of works.
>
> I watch and in its work of creation nature brings forth all that moves and moves not: and thus the revolutions of the world go round.[10]

The Hindu sages were not afraid to identify this rhythmic divine play with the evolution of the cosmos as a whole. They pictured the universe as periodically expanding and contracting and gave the name *kalpa* to the unimaginable time span between the beginning and the end of one creation. The scale of this ancient myth is indeed staggering; it has taken the human mind more than two thousand years to come up again with a similar concept.

From the world of the very large, from the expanding cosmos, let us now return to the world of the infinitely small. Physics in the twentieth century has been characterized by an ever-progressing penetration into this world of submicroscopic dimensions, down into the

* See p. 77.

realms of atoms, nuclei, and their constituents. This exploration of the submicroscopic world has been motivated by one basic question which has occupied and stimulated human thought throughout the ages: what is matter made of? Ever since the beginning of natural philosophy, man has speculated about this question, trying to find the "basic stuff" of which all matter is made; but only in our century has it been possible to seek an answer by undertaking experiments. With the help of a highly sophisticated technology, physicists were able to explore first the structure of atoms, finding that they consisted of nuclei and electrons, and then the structure of the atomic nuclei which were found to consist of protons and neutrons, commonly called nucleons. In the last two decades, they have gone yet another step farther and have begun to investigate the structure of the nucleons—the constituents of the atomic nuclei—which, again, do not seem to be the ultimate elementary particles, but seem to be composed of other entities.

The first step in the penetration into ever-deeper layers of matter—the exploration of the world of atoms—has led to several profound modifications of our view of matter which have been discussed in the previous chapters. The second step was the penetration of the world of atomic nuclei and their constituents, and it has forced us to change our views in a way which is no less profound. In this world, we deal with dimensions which are a hundred thousand times smaller than atomic dimensions, and consequently the particles confined to such small dimensions move considerably faster than those confined to atomic structures. They move, in fact, so fast that they can only be described adequately in the framework of the special theory of relativity. To understand the properties and interactions of subatomic particles, it is thus necessary to use a framework which takes into account both quantum theory and relativity theory, and it is relativity theory which forces us to modify our view of matter once more.

The characteristic feature of the relativistic framework is, as mentioned previously, that it unifies basic concepts which seemed totally unrelated before. One of the most important examples is the equivalence of mass and energy which is expressed mathematically by Einstein's famous equation $E = mc^2$. To understand the profound significance of this equivalence, we first have to understand the meaning of energy, and the meaning of mass.

Energy is one of the most important concepts used in the description of natural phenomena. As in everyday life, we say that a body has energy when it has the capacity for doing work. This energy can take a great variety of forms. It can be energy of motion, energy of heat, gravitational energy, electrical energy, chemical energy, and so on. Whatever the form is, it can be used to do work. A stone, for example, can be given gravitational energy by lifting it up to some height. When it is dropped from that height, its gravitational energy is transformed into energy of motion (kinetic energy), and when the stone hits the ground it can do work by breaking something. Taking a more constructive example, electrical energy or chemical energy can be transformed into heat energy and used for domestic purposes. In physics, energy is always associated with some process, or some kind of activity, and its fundamental importance lies in the fact that the total energy involved in a process is always conserved. It may change its form in the most complicated way, but none of it can get lost. The conservation of energy is one of the most fundamental laws of physics. It governs all known natural phenomena and no violation of the law has so far been observed.

The mass of a body, on the other hand, is a measure of its own weight: i.e., of the pull of gravity on the body. Besides that, mass measures the inertia of an object: i.e., its resistance against being accelerated. Heavy objects are harder to accelerate than light objects, a fact which is well known to anybody who has ever pushed a car. In classical physics, mass was furthermore associated with an indestructible material substance: i.e., with the "stuff" of which all things were thought to be made. Like energy, it was believed to be rigorously conserved, so that no mass could ever get lost.

Now relativity theory tells us that mass is nothing but a form of energy. Energy can not only take the various forms known in classical physics, but can also be locked up in the mass of an object. The amount of energy contained, for example, in a particle is equal to the particle's mass, m, times c^2, the square of the speed of light; thus:

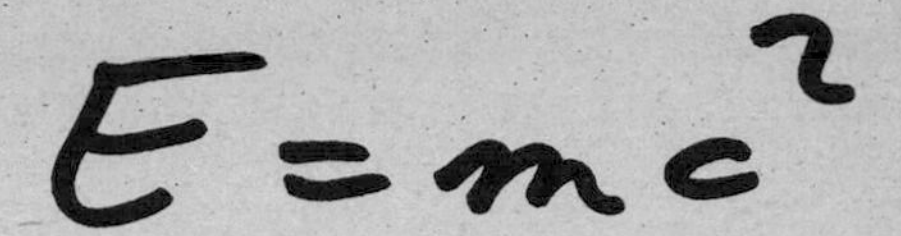

Once it is seen to be a form of energy, mass is no longer required to be indestructible, but can be transformed into other forms of energy. This can happen when subatomic particles collide with one another. In such collisions, particles can be destroyed, and the energy contained in their masses can be transformed into kinetic energy and distributed among the other particles participating in the collision. Conversely, when particles collide with very high velocities, their kinetic energy can be used to form the masses of new particles. The photograph below shows an extreme example of such

a collision: a proton enters the bubble chamber from the left, knocks an electron out of an atom (spiral track), and then collides with another proton to create sixteen new particles in the collision process.

The creation and destruction of material particles is one of the most impressive consequences of the equivalence of mass and energy. In the collision processes of high-energy physics, mass is no longer conserved. The colliding particles can be destroyed and their masses may be transformed partly into the masses, and partly into the kinetic energies of the newly created particles. Only the total energy involved in such a process, that is, the total kinetic energy plus the energy contained in all the masses, is conserved. The collisions of subatomic particles are our main tool to study their properties and the relation between mass and energy is essential for their description. It has been verified innumerable times and particle physicists are completely familiar with the equivalence of mass and energy; so familiar, in fact, that they measure the masses of particles in the corresponding energy units.

The discovery that mass is nothing but a form of energy has forced us to modify our concept of a particle in an essential way. In modern physics, mass is no longer associated with a material substance, and hence particles are not seen as consisting of any basic "stuff," but as bundles of energy. Since energy, however, is associated with activity, with processes, the implication is that the nature of subatomic particles is intrinsically dynamic. To understand this better, we must remember that these particles can only be conceived in relativistic terms, that is, in terms of a framework where space and time are fused into a four-dimensional continuum. The particles must not be pictured as static three-dimensional objects, like billiard balls or grains of sand, but rather as four-dimensional entities in space-time. Their forms have to be understood dynamically, as forms in space and time. Subatomic particles are dynamic patterns which have a space aspect and a time aspect. Their space aspect makes them appear as objects with a certain mass, their time aspect as processes involving the equivalent energy.

These dynamic patterns, or "energy bundles," form the stable nuclear, atomic, and molecular structures which build up matter and give it its macroscopic solid aspect, thus making us believe that it is made of some material substance. At the macroscopic level, this notion of substance is a useful approximation, but at the atomic level it no longer makes sense. Atoms consist of particles, and these particles are not made of any material stuff. When we observe them, we never see any substance; what we observe are dynamic patterns continually changing into one another—a continuous dance of energy.

Quantum theory has shown that particles are not isolated grains of matter, but are probability patterns, interconnections in an inseparable cosmic web. Relativity theory, so to speak, has made these patterns come alive by revealing their intrinsically dynamic character. It has shown that the activity of matter is the very essence of its being. The particles of the subatomic world are not only active in the sense of moving around very fast; they themselves are processes! The existence of matter and its activity cannot be separated. They are but different aspects of the same space-time reality.

It has been argued in the previous chapter that the awareness of the "interpenetration" of space and time has led the Eastern mystics to an intrinsically dynamic world-view. A study of their writings reveals that they conceive the world not only in terms of movement, flow and change, but also seem to have a strong intuition for the "space-time" character of material objects which is so typical of relativistic physics. Physicists have to take into account the unification of space and time when they study the subatomic world and, consequently, they view the objects of this world—the particles—not statically, but dynamically, in terms of energy, activity and processes. The Eastern mystics, in their nonordinary states of consciousness, seem to be aware of the interpenetration of space and time at a macroscopic level, and thus they see the macroscopic objects in a way which is very similar to the physicists' conception of subatomic particles. This is particularly striking in Buddhism. One of the principal teachings of the Buddha was that "all compounded things are impermanent." In the original Pali version of this famous saying,[11] the term used for "things" is *sankhara* (Sanskrit: *samskara*), a word which means first of all "an event" or "a happening"—also "a deed," "an act"—and only secondarily "an existing thing." This clearly shows that Buddhists have a dynamic conception of things as ever-changing processes. In the words of D.T. Suzuki:

> Buddhists have conceived an object as an event and not as a thing or substance. . . . The Buddhist conception of 'things' as *samskara* (or *sankhara*), that is, as 'deeds,' or 'events,' makes it clear that Buddhists understand our experience in terms of time and movement.[12]

Like modern physicists, Buddhists see all objects as processes in a universal flux and deny the existence of any material substance. This denial is one of the most characteristic features of all schools of Buddhist philosophy. It is also characteristic of Chinese thought which developed a similar view of things as transitory stages in the ever-flowing *Tao* and was more concerned with their interrelations than with their reduction to a fundamental substance. "While European philosophy tended to find reality in substance," writes Joseph Needham, "Chinese philosophy tended to find it in relation."[13]

In the dynamic world-views of Eastern mysticism and of modern physics, then, there is no place for static shapes, or for any material substance. The basic elements of the universe are dynamic patterns; transitory stages in the "constant flow of transformation and change," as Chuang Tzu calls it.

According to our present knowledge of matter, its basic patterns are the subatomic particles, and the understanding of their properties and interactions is the principal aim of modern fundamental physics. We know today over two hundred particles, most of them being created artificially in collision processes and living only an extremely short time; far less than a millionth of a second! It is thus quite obvious that these short-lived particles represent merely transitory patterns of dynamic processes. The main questions with regard to these patterns, or particles, are the following: What are their distinguishing features? Are they composite and, if so, what do they consist of, or—better—what other patterns do they involve? How do they interact with one another: i.e., what are the forces between them? Last, if the particles themselves are processes, what kind of processes are they?

We have become aware that in particle physics all these questions are inseparably connected. Because of the relativistic nature of subatomic particles, we cannot understand their properties without understanding their mutual interactions, and because of the basic interconnectedness of the subatomic world we shall not understand any one particle before understanding all the others. The following chapters will show how far we have come in understanding the particles' properties and interactions.

Although we are still lacking a complete quantum-relativistic theory of the subatomic world, several partial theories and models have been developed which describe some aspects of this world very successfully. A discussion of the most important of these models and theories will show that they all involve philosophical conceptions which are in striking agreement with those in Eastern mysticism.

14
EMPTINESS AND FORM

The classical mechanistic world-view was based on the notion of solid, indestructible particles moving in the void. Modern physics has brought about a radical revision of this picture. It has led not only to a completely new notion of "particles," but has also transformed the classical concept of the void in a profound way. This transformation took place in the so-called field theories. It began with Einstein's idea of associating the gravitational field with the geometry of space and became even more pronounced when quantum theory and relativity theory were combined to describe the force fields of subatomic particles. In these "quantum field theories," the distinction between particles and the space surrounding them loses its original sharpness, and the void is recognized as a dynamic quantity of paramount importance.

The field concept was introduced in the nineteenth century by Faraday and Maxwell in their description of the forces between electric charges and currents. An electric field is a condition in the space around a charged body which will produce a force on any other charge in that space. Electric fields are thus created by charged bodies and their effects can only be felt by charged bodies. Magnetic fields are produced by charges in motion, i.e., by electric currents, and the magnetic forces resulting from them can be felt by other moving charges. In classical electrodynamics, the theory constructed by Faraday and Maxwell, the fields are primary physical entities which can be studied without any reference to material bodies. Vibrating electric and magnetic fields can travel through space in the form of radio waves, light waves, or other kinds of electromagnetic radiation.

Relativity theory has made the structure of electrodynamics much more elegant by unifying the concepts of both charges and currents and electric and magnetic fields. Since all motion is relative, every charge can also appear as a current—in a frame of reference where it moves with respect to the observer—and consequently, its electric field can also appear as a magnetic field. In the relativistic formulation of electrodynamics, the two fields are thus unified into a single electromagnetic field.

The concept of a field has been associated not only with the electromagnetic force, but also with that other major force in the large-scale world, the force of gravity. Gravitational fields are created and felt by all massive bodies, and the resulting forces are always forces of attraction, contrary to the electromagnetic fields which are felt only by charged bodies and which give rise to attractive and repulsive forces. The proper field theory for the gravitational field is the general theory of relativity, and in this theory the influence of a massive body on the surrounding space is more far-reaching than the corresponding influence of a charged body in electrodynamics. Again, the space around the object is "conditioned" in such a way that another object will feel a force, but this time the conditioning affects the geometry, and thus the very structure of space.

Matter and empty space—the full and the void—were the two fundamentally distinct concepts on which the atomism of Democritus and of Newton was based. In general relativity, these two concepts can no longer be separated. Wherever there is a massive body, there will also be a gravitational field, and this field will manifest itself as the curvature of the space surrounding that body. We must not think, however, that the field fills the space and "curves" it. The two cannot be distinguished; the field *is* the curved space! In general relativity, the gravitational field and the structure—or geometry—of space are identical. They are represented in Einstein's field equations by one and the same mathematical quantity. In Einstein's theory, then, matter cannot be separated from its field of gravity, and the field of gravity cannot be separated from the curved space.

Matter and space are thus seen to be inseparable and interdependent parts of a single whole.

Material objects not only determine the structure of the surrounding space but are, in turn, influenced by their environment in an essential way. According to the physicist and philosopher Ernst Mach, the inertia of a material object—the object's resistance against being accelerated—is not an intrinsic property of matter, but a measure of its interaction with all the rest of the universe. In Mach's view, matter has inertia only because there is other matter in the universe. When a body rotates, its inertia produces centrifugal forces (used, for example, in a spin-drier to extract water from wet laundry), but these forces appear only because the body rotates "relative to the fixed stars," as Mach has put it. If those fixed stars were suddenly to disappear, the inertia and the centrifugal forces of the rotating body would disappear with them.

This conception of inertia, which has become known as Mach's principle, had a deep influence on Albert Einstein and was his original motivation for constructing the general theory of relativity. Due to the considerable mathematical complexity of Einstein's theory, physicists have not yet been able to agree whether it actually incorporates Mach's principle or not. Most physicists believe, however, that it should be incorporated, in one way or another, into a complete theory of gravity.

Thus modern physics shows us once again—and this time at the macroscopic level—that material objects are not distinct entities, but are inseparably linked to their environment; that their properties can only be understood in terms of their interaction with the rest of the world. According to Mach's principle, this interaction reaches out to the universe at large, to the distant stars and galaxies. The basic unity of the cosmos manifests itself, therefore, not only in the world of the very small but also in the world of the very large; a fact which is increasingly acknowledged in modern astrophysics and cosmology. In the words of the astronomer Fred Hoyle:

> Present-day developments in cosmology are coming to suggest rather insistently that everyday conditions could not persist but for the distant parts of the

> Universe, that all our ideas of space and geometry would become entirely invalid if the distant parts of the Universe were taken away. Our everyday experience even down to the smallest details seems to be so closely integrated to the grand-scale features of the Universe that it is well-nigh impossible to contemplate the two being separated.[1]

The unity and interrelation between a material object and its environment, which is manifest on the macroscopic scale in the general theory of relativity, appears in an even more striking form at the subatomic level. Here, the ideas of classical field theory are combined with those of quantum theory to describe the interactions between subatomic particles. Such a combination has not yet been possible for the gravitational interaction because of the complicated mathematical form of Einstein's theory of gravity; but the other classical field theory, electrodynamics, has been merged with quantum theory into a theory called "quantum electrodynamics" which describes all electromagnetic interactions between subatomic particles. This theory incorporates both quantum theory and relativity theory. It was the first "quantum-relativistic" model of modern physics and is still the most successful.

The striking new feature of quantum electrodynamics arises from the combination of two concepts: that of the electromagnetic field, and that of photons as the particle manifestations of electromagnetic waves. Since photons are also electromagnetic waves, and since these waves are vibrating fields, the photons must be manifestations of electromagnetic fields. Hence the concept of a "quantum field": that is, of a field which can take the form of quanta, or particles. This is indeed an entirely new concept which has been extended to describe all subatomic particles and their interactions, each type of particle corresponding to a different field. In these "quantum field theories," the classical contrast between the solid particles and the space surrounding them is completely overcome. The quantum field is seen as the fundamental physical entity: a continuous medium which is present everywhere in space. Particles are merely local condensations of the field; concentrations of energy which come and go, thereby losing their individual character and dissolving into the underlying field. In the words of Albert Einstein:

> We may therefore regard matter as being constituted by the regions of space in which the field is extremely intense. . . . There is no place in this new kind of physics both for the field and matter, for the field is the only reality.[2]

The conception of physical things and phenomena as transient manifestations of an underlying fundamental entity is not only a basic element of quantum field theory, but also a basic element of the Eastern world view. Like Einstein, the Eastern mystics consider this underlying entity as the only reality: all its phenomenal manifestations are seen as transitory and illusory. This reality of the Eastern mystic cannot be identified with the quantum field of the physicist because it is seen as the essence of *all* phenomena in this world and, consequently, is beyond all concepts and ideas. The quantum field, on the other hand, is a well-defined concept which only accounts for some of the physical phenomena. Nevertheless, the intuition behind the physicist's interpretation of the subatomic world, in terms of the quantum field, is closely paralleled by that of the Eastern mystic who interprets his or her experience of the world in terms of an ultimate underlying reality. Subsequent to the emergence of the field concept, physicists have attempted to unify the various fields into a single fundamental field which would incorporate all physical phenomena. Einstein, in particular, spent the last years of his life searching for such a unified field. The *Brahman* of the Hindus, like the *Dharmakaya* of the Buddhists and the *Tao* of the Taoists, can be seen, perhaps, as the ultimate unified field from which spring not only the phenomena studied in physics, but all other phenomena as well.

In the Eastern view, the reality underlying all phenomena is beyond all forms and defies all description and specification. It is therefore often said to be formless, empty or void. But this emptiness is not to be taken for mere nothingness. It is, on the contrary, the essence of all forms and the source of all life. Thus the *Upanishads* say,

> *Brahman* is life. *Brahman* is joy. *Brahman* is the Void. . . .
> Joy, verily, that is the same as the Void.
> The Void, verily, that is the same as joy.[3]

Buddhists express the same idea when they call the ultimate reality *Sunyata*—'Emptiness,' or 'the Void'—and affirm that it is a living Void which gives birth to all forms in the phenomenal world. The Taoists ascribe a similar infinite and endless creativity to the *Tao* and, again, call it empty. "The *Tao* of Heaven is empty and formless," says the *Kuan-tzu*,[4] and Lao Tzu uses several metaphors to illustrate this emptiness. He often compares the *Tao* to a hollow valley, or to a vessel which is forever empty and thus has the potential of containing an infinity of things.

In spite of using terms like 'empty' and 'void,' the Eastern sages make it clear that they do not mean ordinary emptiness when they talk about *Brahman, Sunyata* or *Tao*, but, on the contrary, a Void which has an infinite creative potential. Thus, the Void of the Eastern mystics can easily be compared to the quantum field of subatomic physics. Like the quantum field, it gives birth to an infinite variety of forms which it sustains and, eventually, reabsorbs. As the *Upanishads* say:

> Tranquil, let one worship It
> As that from which he came forth,
> As that into which he will be dissolved,
> As that in which he breathes.[5]

The phenomenal manifestations of the mystical Void, like the subatomic particles, are not static and permanent, but dynamic and transitory, coming into being and vanishing in one ceaseless dance of movement and energy. Like the subatomic world of the physicist, the phenomenal world of the Eastern mystic is a world of *samsara*—of continuous birth and death. Being transient manifestations of the Void, the things in this world do not have any fundamental identity. This is especially emphasized in Buddhist philosophy which denies the existence of any material substance and also holds that the idea of a constant 'self' undergoing successive experiences is an illusion. Buddhists have frequently compared this illusion of a material substance and an individual self to the phenomenon of a water wave, in which the up-and-down movement of the water particles makes us believe that a "piece" of water moves over the surface.* It is interesting to note that physicists have used the same analogy in the context of field theory to point out the illusion of a material substance created by a moving particle. Thus Hermann Weyl writes:

> According to the [field theory of matter] a material particle such as an electron is merely a small domain of the electrical field within which the field strength assumes enormously high values, indicating that a comparatively huge field energy is concentrated in a very small space. Such an energy knot, which by no means is clearly delineated against the remaining field, propagates through empty space like a water wave across the surface of a lake; there is no such thing as one and the same substance of which the electron consists at all times.[6]

In Chinese philosophy, the field idea is not only implicit in the notion of the *Tao* as being empty and formless, and yet producing all forms, but is also expressed explicitly in the concept of *ch'i*. This term played an important role in almost every Chinese school of natural philosophy and was particularly important in Neo-Confucianism; the school which attempted a synthesis of Confucianism, Buddhism, and Taoism.** The word *ch'i* literally means "gas" or "ether," and was used in ancient China to denote the vital breath or energy animating the cosmos. In the human body, the "pathways of *ch'i*" are the basis of traditional Chinese medicine.

* See p. 137.
** See p. 92.

The aim of acupuncture is to stimulate the flow of *ch'i* through these channels. The flow of *ch'i* is also the basis of the flowing movements of *T'ai Chi Ch'uan*, the Taoist dance of the warrior.

The Neo-Confucians developed a notion of *ch'i* which bears the most striking resemblance to the concept of the quantum field in modern physics. Like the quantum field, *ch'i* is conceived as a tenuous and nonperceptible form of matter which is present throughout space and can condense into solid material objects. In the words of Chang Tsai:

> When the *ch'i* condenses, its visibility becomes apparent so that there are then the shapes [of individual things]. When it disperses, its visibility is no longer apparent and there are no shapes. At the time of its condensation, can one say otherwise than that this is but temporary? But at the time of is dispersing, can one hastily say that it is then nonexistent?[7]

Thus *ch'i* condenses and disperses rhythmically, bringing forth all forms which eventually dissolve into the Void. As Chang Tsai says again:

> The Great Void cannot but consist of *ch'i*; this *chi'i* cannot but condense to form all things; and these things cannot but become dispersed so as to form [once more] the Great Void.[8]

As in quantum field theory, the field—or the *ch'i*—is not only the underlying essence of all material objects, but also carries their mutual interactions in the form of waves. The following descriptions of the field concept in modern physics by Walter Thirring, and of the Chinese view of the physical world by Joseph Needham, make the strong similarity apparent:

> Modern theoretical physics . . . has put our thinking about the essence of matter in a different context. It has taken our gaze from the visible—the particles —to the underlying entity, the field. The presence of matter is merely a disturbance of the perfect state of the field at that place; something accidental, one

> could almost say, merely a "blemish." Accordingly, there are no simple laws describing the forces between elementary particles. . . . Order and symmetry must be sought in the underlying field.[9]

> The Chinese physical universe in ancient and medieval times was a perfectly continuous whole. *Ch'i* condensed in palpable matter was not particulate in any important sense, but individual objects acted and reacted with all other objects in the world . . . in a wavelike or vibratory manner dependent, in the last resort, on the rhythmic alternation at all levels of the two fundamental forces, the *yin* and the *yang*. Individual objects thus had their intrinsic rhythms. And these were integrated . . . into the general pattern of the harmony of the world.[10]

With the concept of the quantum field, modern physics has found an unexpected answer to the old question of whether matter consists of indivisible atoms or of an underlying continuum. The field is a continuum which is present everywhere in space and yet in its particle aspect has a discontinuous, "granular" structure. The two apparently contradictory concepts are thus unified and seen to be merely different aspects of the same reality. As always in a relativistic theory, the unification of the two opposite concepts takes place in a dynamic way: the two aspects of matter transform themselves endlessly into one another. Eastern mysticism emphasizes a similar dynamic unity between the Void and the forms which it creates. In the words of Lama Govinda:

> The relationship of form and emptiness cannot be conceived as a state of mutually exclusive opposites, but only as two aspects of the same reality, which coexist and are in continual cooperation.[11]

The fusion of these opposite concepts into a single whole has been expressed in a Buddhist *sutra* in the celebrated words:

> Form is emptiness, and emptiness is indeed form. Emptiness is not different from form; form is not

different from emptiness. What is form that is emptiness; what is emptiness that is form.[12]

The field theories of modern physics have led not only to a new view of subatomic particles but have also decisively modified our notions about the forces between these particles. The field concept was originally linked to the concept of force, and even in quantum field theory it is still associated with the forces between particles. The electromagnetic field, for example, can manifest itself as a "free field" in the form of traveling waves/photons, or it can play the role of a field of force between charged particles. In the latter case, the force manifests itself as the exchange of photons between the interacting particles. The electric repulsion between two electrons, for example, is mediated through these photon exchanges.

This new notion of a force may seem difficult to understand, but it becomes much clearer when the process of exchanging a photon is pictured in a space-time diagram. The diagram below shows two electrons

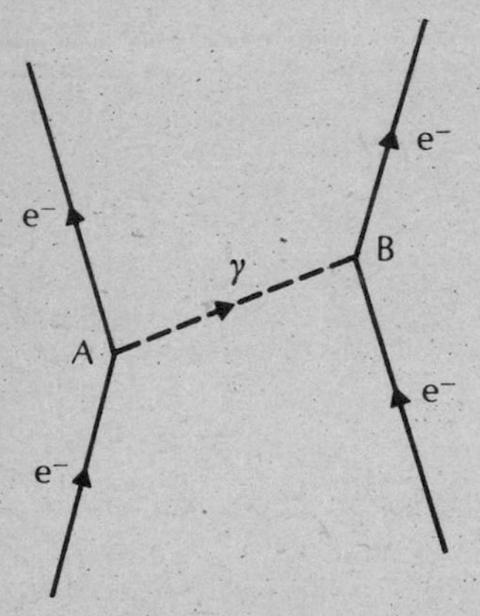

mutual repulsion of two electrons through the exchange of a photon

approaching each other, one of them emitting the photon (denoted by γ) at the point A, the other one absorbing it at the point B. When the first electron emits the photon it reverses its direction and changes its velocity (as can be seen from the different direction and inclination of its world line), and so does the second electron when it absorbs the photon. In the end, the two electrons fly apart, having repelled each other through the exchange of the photon. The full interaction between the electrons will involve a series of photon exchanges, and as a result the electrons will appear to deflect one another along smooth curves.

In terms of classical physics, one would say that the electrons exert a repulsive force on one another. This, however, is now seen to be a very imprecise way of describing the situation. Neither of the two electrons "feels" a force when they approach each other. All they do is interact with the exchanged photons. The force is nothing but the collective macroscopic effect of these multiple photon exchanges. The concept of force is therefore no longer useful in subatomic physics. It is a classical concept which we associate (even if only subconsciously) with the Newtonian idea of a force being felt over a distance. In the subatomic world there are no such forces, but only interactions between particles, mediated through fields, that is, through other particles. Hence physicists prefer to speak about interactions, rather than about forces.

According to quantum field theory, all interactions take place through the exchange of particles. In the case of electromagnetic interactions, the exchanged particles are photons; nucleons, on the other hand, interact through the much stronger nuclear force—or "strong interaction" —which manifests itself as the exchange of a new kind of particles called "mesons." There are many different types of mesons which can be exchanged between protons and neutrons. The closer the nucleons are to each other, the more numerous and heavy the mesons they exchange. The interactions between nucleons are thus linked to the properties of the exchanged mesons and these, in turn, interact mutually through the exchange of other particles. For this reason, we shall not be able to understand the nuclear force on a fundamental level without understanding the whole spectrum of subatomic particles.

In quantum field theory, all particle interactions can be pictured in space-time diagrams, and each diagram is associated with a mathematical expression which allows one to calculate the probability for the corresponding process to occur. The exact correspondence between the diagrams and the mathematical expressions was established in 1949 by Richard Feynman, since when the diagrams have been known as Feynman diagrams. A crucial feature of the theory is the creation and destruction of particles. For example, the photon in our diagram is created in the process of emission at point A, and is destroyed when it is absorbed at point B. Such a process can be conceived in a relativistic theory only where particles are not seen as indestructible objects, but rather as dynamic patterns involving a certain amount of energy which can be redistributed when new patterns are formed.

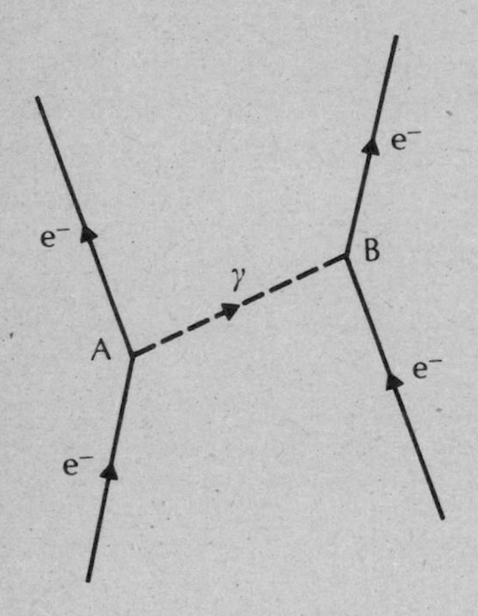

The creation of a massive particle is only possible when the energy corresponding to its mass is provided, for example, in a collision process. In the case of the strong interactions, this energy is not always available, as when two nucleons interact with one another in an atomic nucleus. In such cases, the exchange of massive mesons should therefore not be possible. Yet, these exchanges do take place. Two protons, for example, may exchange a "pi-meson," or "pion," whose mass is about one-seventh of the proton mass:

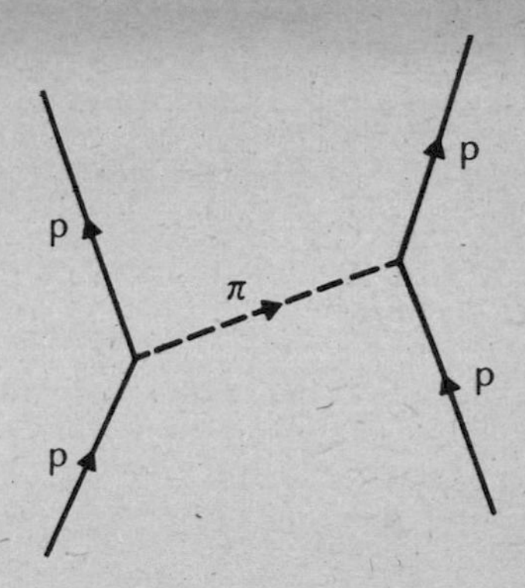

exchange of a pion (π)
between two protons (p)

The reasons why exchange processes of that kind can happen, in spite of the apparent lack of energy for creating the meson, is to be found in a "quantum effect" connected with the uncertainty principle. As discussed previously,* subatomic events ocurring within a short time span involve a large uncertainty of energy. Exchanges of mesons—i.e., their creation and subsequent destruction—are events of that kind. They take place during such a short time that the uncertainty of energy is enough to allow for the creation of the mesons. These mesons are called "virtual" particles. They are different from the "real" mesons created in collision processes because they can exist only during the period of time allowed by the uncertainty principle. The heavier the mesons are (i.e., the more energy is required to create them), the shorter is the time allowed for the exchange process. This is why nucleons can exchange heavy mesons only when they are very close together. The exchange of virtual photons, on the other hand, can take place over indefinite distances because the photons, being massless, can be created with indefinitely small amounts of energy. This analysis of nuclear and electromagnetic forces enabled Hideki Yukawa in 1935 not only to predict the existence of the pion twelve years before it was observed,

* See p. 144.

but also approximately to estimate its mass from the range of the nuclear force.

In quantum field theory, then, all interactions are pictured as the exchange of virtual particles. The stronger the interaction—i.e., the stronger the resulting "force" between the particles—the higher the probability of such exchange processes; the more frequently will virtual particles be exchanged. The role of virtual particles, however, is not limited to these interactions. One nucleon alone, for example, may very well emit a virtual particle and reabsorb it shortly afterward. Provided the created meson disappears within the time allowed by the uncertainty principle, there is nothing to forbid such a process. The corresponding Feynman diagram for a neutron emitting and reabsorbing a pion is reproduced below.

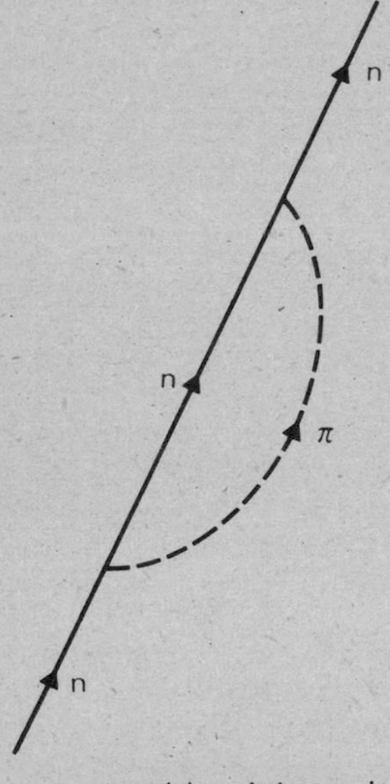

a neutron (n) emitting and reabsorbing a pion

The probability for such "self-interaction" processes is very high for nucleons because of their strong interaction. This means that nucleons are, in fact, emitting and absorbing virtual particles all the time. According to field theory, they have to be regarded as centers of continuous activity surrounded by clouds of virtual particles. The virtual mesons have to disappear very shortly after their creation, which means they cannot move very far away from the nucleon. The meson cloud is thus very small. Its outer regions are populated by light mesons (mostly pions), the heavier mesons having to be absorbed after a much shorter time and therefore being confined to the inner parts of the cloud.

Every nucleon is surrounded by such a cloud of virtual mesons which live only for an exceedingly short period of time. However, virtual mesons may become real mesons under special circumstances. When a nucleon is hit by another particle moving with a high velocity, some of the energy of motion of that particle may be transferred to a virtual meson to free it from the cloud. This is how real mesons are created in high-energy collisions. On the other hand, when two nucleons come so near to each other that their meson clouds overlap, some of the virtual particles may not go back to be absorbed by the nucleon which originally created them, but may "jump across," to be absorbed by the other nucleon. This is how the exchange processes arise which constitute the strong interactions.

This picture shows clearly that the interactions between particles, and thus the "forces" between them, are determined by the composition of their virtual clouds. The range of an interaction—that is, the distance between the particles at which the interaction will set in—depends on the extension of the virtual clouds, and the detailed form of the interaction will depend on the properties of the particles present in the clouds. Thus the electromagnetic forces are due to the presence of virtual photons "within" charged particles, whereas the strong interactions between nucleons arise from the presence of virtual pions and other mesons "within" the nucleons. In field theory, the forces between particles appear as intrinsic properties of the particles. Force and matter, the two concepts that were so sharply separated in Greek and Newtonian atomism, are now seen to have their common origin in the dynamic patterns that we call particles.

Such a view of forces is also characteristic of Eastern mysticism, which regards motion and change as essential and intrinsic properties of all things. "All rotating things," says Chang Tsai with reference to the heavens, "have a spontaneous force and thus their motion is not imposed on them from outside";[13] and in the *I Ching* we read:

> [The natural] laws are not forces external to things, but represent the harmony of movement immanent in them.[14]

This ancient Chinese description of forces as representing the harmony of movement within things seems particularly appropriate in the light of quantum field theory, where the forces between particles are seen as reflecting dynamic patterns (the virtual clouds) inherent in these particles.

The field theories of modern physics force us to abandon the classical distinction between material particles and the void. Einstein's field theory of gravity and quantum field theory both show that particles cannot be separated from the space surrounding them. On the one hand, they determine the structure of that space, while on the other hand they cannot be regarded as isolated entities, but have to be seen as condensations of a continuous field which is present throughout space. In quantum field theory, this field is seen as the basis of all particles and of their mutual interactions.

> The field exists always and everywhere; it can never be removed. It is the carrier of all material phenomena. It is the "void" out of which the proton creates the pi-mesons. Being and fading of particles are merely forms of motion of the field.[15]

The distinction between matter and empty space finally had to be abandoned when it became evident that virtual particles can come into being spontaneously out of the void, and vanish again into the void, without any nucleon or other strongly interacting particle being present. Here is a "vacuum diagram" for such a process: three particles —a proton (p), an antiproton ($\bar{p}$), and a pion (π)—are formed out of nothing and disappear again into the

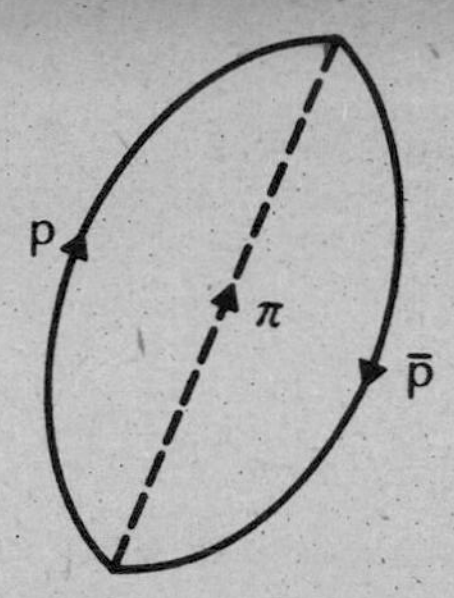

a vacuum diagram

vacuum. According to field theory, events of that kind happen all the time. The vacuum is far from empty. On the contrary, it contains an unlimited number of particles which come into being and vanish without end.

Here then, is the closest parallel to the Void of Eastern mysticism in modern physics. Like the Eastern Void, the "physical vacuum"—as it is called in field theory—is not a state of mere nothingness, but contains the potentiality for all forms of the particle world. These forms, in turn, are not independent physical entities, but merely transient manifestations of the underlying Void. As the *sutra* says, "Form is emptiness, and emptiness is indeed form."

The relation between the virtual particles and the vacuum is an essentially dynamic relation; the vacuum is truly a "living Void," pulsating in endless rhythms of creation and destruction. The discovery of the dynamic quality of the vacuum is seen by many physicists as one of the most important findings of modern physics. From its role as an empty container of the physical phenomena, the void has emerged as a dynamic quantity of utmost importance. The results of modern physics thus seem to confirm the words of the Chinese sage Chang Tsai:

> When one knows that the Great Void is full of *ch'i*, one realizes that there is no such thing as nothingness.[16]

15
THE COSMIC DANCE

The exploration of the subatomic world in the twentieth century has revealed the intrinsically dynamic nature of matter. It has shown that the constituents of atoms, the subatomic particles, are dynamic patterns which do not exist as isolated entities, but as integral parts of an inseparable network of interactions. These interactions involve a ceaseless flow of energy manifesting itself as the exchange of particles; a dynamic interplay in which particles are created and destroyed without end in a continual variation of energy patterns. The particle interactions give rise to the stable structures which build up the material world, which again do not remain static, but oscillate in rhythmic movements. The whole universe is thus engaged in endless motion and activity; in a continual cosmic dance of energy.

This dance involves an enormous variety of patterns, but, surprisingly, they fall into a few distinct categories. The study of the subatomic particles and their interactions thus reveals a great deal of order. All atoms, and consequently all forms of matter in our environment, are composed of only three massive particles: the proton, the neutron, and the electron. A fourth particle, the photon, is massless and represents the unit of electromagnetic radiation. The proton, the electron, and the photon are all stable particles, which means they live forever unless they become involved in a collision process where they can be annihilated. The neutron, on the other hand, can disintegrate spontaneously. This disintegration is called "beta decay" and is the basic process of a certain type of radioactivity. It involves the transformation of the neutron into a proton, accompanied by the creation of an electron and a new type of massless particle, called the neutrino. Like the proton and the electron, the neutrino is also stable. It is commonly denoted by the Greek letter ν ("nu"), and the process of beta decay is symbolically written as

$$n \rightarrow p + e^- + \nu$$

The transformation of neutrons into protons in the atoms of a radioactive substance entails a transformation of these atoms into atoms of an entirely different kind. The electrons which are created in the process are emitted as a powerful radiation which is widely used in biology, medicine, and industry. The neutrinos, on the other hand, although emitted in equal number, are very difficult to detect because they have neither mass nor electric charge.

As mentioned previously, there is an antiparticle for every particle, with equal mass but opposite charge. The photon is its own antiparticle; the antiparticle of the electron is called the positron; then there is an antiproton, an antineutron, and an antineutrino. The massless particle created in beta decay is not, in fact, the neutrino but the antineutrino (denoted by $\bar{\nu}$), so that the process is correctly written as

$$n \rightarrow p + e^- + \bar{\nu}$$

The particles mentioned so far represent only a fraction of the subatomic particles known today. All the others are unstable and decay after a very short time into other particles, some of which may decay again until a combination of stable particles remains. The study of unstable particles is very expensive as they have to be newly created in collision processes for each investigation, which involves huge particle accelerators, bubble chambers, and other extremely sophisticated devices for particle detection.

Most unstable particles live only for an extremely short time, compared with the human time scale: less than a millionth of a second. However, their lifetime has to be regarded in relation to their size, which is also diminutive. When looked at in this way, it can be seen that many of them live for a relatively long period, and that

		NAME	SYMBOL	
			PARTICLE	ANTIPARTICLE
		photon	γ	
leptons		neutrino	ν_e ν_μ	$\bar{\nu}_e$ $\bar{\nu}_\mu$
leptons		electron	e^-	e^+
leptons		muon	μ^-	μ^+
hadrons	mesons	pion	π^+ π^0 π^-	
hadrons	mesons	kaon	K^+ K^0	$\bar{K}^0$ K^-
hadrons	mesons	eta	η	
hadrons	baryons	proton	p	$\bar{p}$
hadrons	baryons	neutron	n	$\bar{n}$
hadrons	baryons	lambda	Λ	$\bar{\Lambda}$
hadrons	baryons	sigma	Σ^+ Σ^0 Σ^-	$\bar{\Sigma}^+$ $\bar{\Sigma}^0$ $\bar{\Sigma}^-$
hadrons	baryons	cascade	Ξ^0 Ξ^-	$\bar{\Xi}^0$ $\bar{\Xi}^-$
hadrons	baryons	omega	Ω	$\bar{\Omega}^-$

The table shows thirteen different types of particles, many of which appear in different "charge states." The pions, for example, can have positive charge (π^+), negative charge (π^-), or be electrically neutral (π^0). There are two kinds of neutrinos, one interacting only with electrons (ν_e), the other only with muons (ν_μ). The antiparticles are listed as well, three of the particles (γ, $\pi\mu$, η), being their own antiparticles. Particles are arranged in the order of increasing mass: the photons and the neutrinos are massless; the electron is the lightest massive particle; the muons, pions, and kaons are a few hundred times heavier than the electron; the other particles are one to three thousand times heavier.

one-millionth of a second is, in fact, an enormous time span in the particle world. A human being can move across a distance a few times his or her size in a second. For a particle, the equivalent time span would therefore be the time it needs to travel over a distance a few times its own size; a unit of time which one could call a "particle second."*

To cross a medium-sized atomic nucleus, a particle needs about ten of these "particle seconds" if it travels at a speed close to the speed of light, as particles do in the collision experiments. Among the great number of unstable particles, there are about two dozen which can travel across at least several atoms before they decay. This is a distance of some 100,000 times their size and corresponds to a time of a few hundred "particle hours." These particles are listed in the table on p. 213, together with the stable particles already mentioned. Most of the unstable particles in the table will, in fact, cover a whole centimeter, or even several centimeters, before they decay, and those which live longest, a millionth of a second, can travel several hundred meters before decaying—an enormous length compared with their size.

All the other particles known so far belong to a category called "resonances" which will be discussed in more detail in the subsequent chapter. They live for a considerably shorter time, decaying after a few "particle seconds," so that they can never travel farther than a few times their size. This means they cannot be seen in the bubble chamber; their existence can only be inferred indirectly. The tracks seen in bubble-chamber pictures can only be traced by particles listed in the table.

All these particles can be created and annihilated in collision processes; each one can also be exchanged as a virtual particle and thus contribute to the interaction between other particles. This would seem to result in a vast number of different particle interactions, but fortunately, although we do not yet know why, all these

* Physicists write this time unit as 10^{-23} seconds which is a shorthand notation for a decimal number with 23 zeroes in front of the figure 1 (including the one in front of the decimal point), i.e., for 0.00000000000000000000001 seconds.

interactions seem to fall into four categories with markedly different interaction strengths:

The strong interactions
The electromagnetic interactions
The weak interactions
The gravitational interactions

Among them, the electromagnetic and gravitational interactions are the most familiar, because they are experienced in the large-scale world. The gravitational interaction acts between all particles, but is so weak it cannot be detected experimentally. In the macroscopic world, however, the huge number of particles making up massive bodies combine their gravitational interaction to produce the force of gravity which is the dominating force in the universe at large. Electromagnetic interactions take place between all charged particles. They are responsible for the chemical processes, and the formation of all atomic and molecular structures. The strong interactions hold the protons and neutrons together in the atomic nucleus. They constitute the nuclear force, by far the strongest of all forces in nature. Electrons, for example, are bound to the atomic nuclei by the electromagnetic force with energies of about ten units (called electron volts), whereas the nuclear force holds protons and neutrons together with energies of about ten million units!

The nucleons are not the only particles interacting through the strong interactions. In fact, the overwhelming majority are strongly interacting particles. Of all the particles known today, only five (and their antiparticles) do not participate in the strong interactions. These are the photon and the four "leptons" listed in the top part of the table. Thus all the particles fall into two broad groups: leptons and "hadrons," or strongly interacting particles. The hadrons are further divided into "mesons" and "baryons" which differ in various ways, one of them being that all baryons have distinct antiparticles, whereas a meson can be its own antiparticle.

The leptons are involved in the fourth type of interactions, the weak interactions. These are so weak, and have such a short range, that they cannot hold anything together, whereas the other three give rise to binding forces—the strong interactions holding together the atomic nuclei, the electromagnetic interactions the atoms and molecules, and the gravitational interactions the planets, stars and galaxies. The weak interactions manifest themselves only in certain kinds of particle collisions and in particle decays, such as the beta decay mentioned earlier.

All interactions between hadrons are mediated by the exchange of other hadrons. It is these exchanges of massive particles that cause the strong interactions to have such a short range.* They extend only over a distance of a few particle sizes and can therefore never build up a macroscopic force. Strong interactions are thus not experienced in the everyday world. The electromagnetic interactions, on the other hand, are mediated by the exchange of massless photons and thus their range is indefinitely long,** which is why the electric and magnetic forces are encountered in the large-scale world. The gravitational interactions, too, are believed to be mediated by a massless particle, called the "graviton," but they are so weak that it has not yet been possible to observe the graviton, although there is no serious reason to doubt its existence.

The weak interactions, finally, have an extremely short range—much shorter than that of the strong interactions—and are therefore assumed to be produced by the exchange of a very heavy particle, called the "W-meson." This hypothetical particle is believed to play a role analogous to that of the photon in the electromagnetic interactions, except for its large mass. This analogy is, in fact, the basis of the most recent developments in field theory in which the formulation of a unified theory of electromagnetic and weak interactions is attempted.

In many of the collision processes of high-energy physics, the strong, electromagnetic, and weak interactions combine to produce an intricate sequence of events. The initial colliding particles are often destroyed and several new particles are created which either un-

* See p. 206.
** See p. 205.

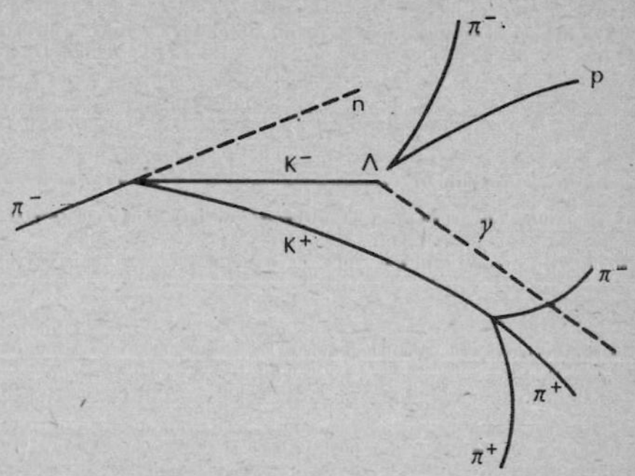

Above
An intricate sequence of particle collisions and decays: a negative pion (π^-), coming in from the left, collides with a proton—i.e., with the nucleus of a hydrogen atom—"sitting" in the bubble chamber; both particles are annihilated, and a neutron (n) plus two kaons (K^- and K^+), are created; the neutron flies off without leaving a track; the K^- collides with another proton in the chamber, the two particles annihilating each other and creating a lambda (Λ) and a photon (γ). Neither of these two neutral particles is visible, but the Λ decays after a very short time into a proton (p) and a π^-, both of which produce tracks. The short distance between the creation of the Λ and its decay can be made out very clearly in the photograph. The K^+, finally, which was created in the initial collision, travels for a while before decaying into three pions.

dergo further collisions or decay, sometimes in several steps, into the stable particles which finally remain. The picture above shows a bubble-chamber photograph* of such a sequence of creation and destruction. It is an impressive illustration of the mutability of matter at the particle level, showing a cascade of energy in which various patterns, or particles, are formed and dissolved.

In these sequences, the creation of matter is particularly striking when a massless, but highly energetic photon, which cannot be seen in the bubble chamber, suddenly explodes into a pair of charged particles—an electron and a positron—sweeping out in divergent curves. Here is a beautiful example of a process involving two of these pair creations.

* Notice that only the charged particles produce tracks in the bubble chamber; these are bent by magnetic fields in a clockwise direction for positively charged particles, and counterclockwise for negative particles.

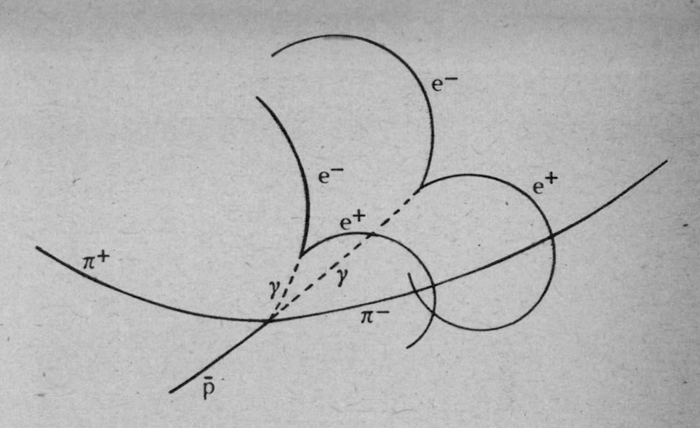

Opposite and above
A sequence of events involving two pair creations: an antiproton ($\bar{p}$), coming from below, collides with one of the protons in the bubble chamber to create a π^+ (flying off to the left), a π^- (flying off to the right), and two photons (γ), each of which creates an electron-positron pair, the positrons (e^+) curving to the right, the electrons (e^-) to the left.

The higher the initial energy in these collision processes, the more particles can be created. The photograph on p. 220 shows the creation of eight pions in a collision between an antlproton and a proton, and the next one is an example of an extreme case; the creation of sixteen particles in a single collision between a pion and a proton.

All these collisions have been produced artificially in the laboratory by the use of huge machines in which the particles are accelerated to the required energies. In most natural phenomena here on earth, the energies are not high enough for massive particles to be created. In outer space, however, the situation is entirely different. Subatomic particles occur in large numbers in the center of the stars where collision processes similar to the ones studied in the accelerator laboratories take place naturally all the time. In some stars, these processes produce an extremely strong electromagnetic radiation—in the form of radio waves, light waves, or X rays—which is the astronomer's primary source of information about the universe. Interstellar space, as well as the space between

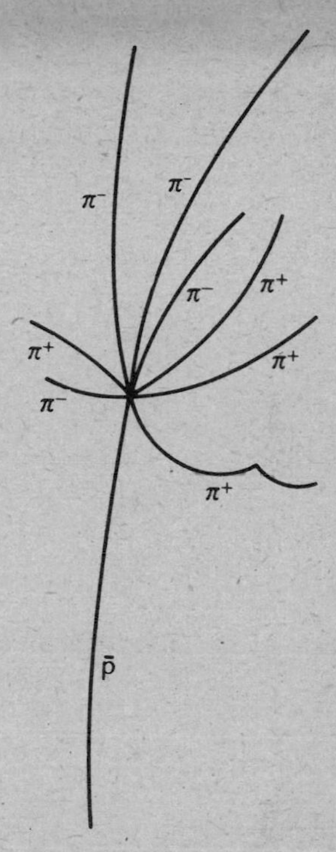

Creation of eight pions in a collision between an antiproton ($\bar{p}$) and a proton (sitting in the bubble chamber); see photograph on opposite page

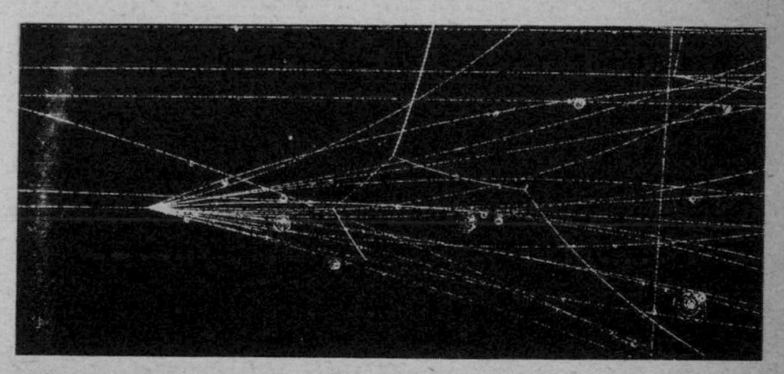

creation of sixteen particles in a pion-proton collision

the galaxies, is thus filled with electromagnetic radiation of various frequencies; i.e., with photons of various energies. These, however, are not the only particles traveling through the cosmos. "Cosmic radiation" contains not only photons but also massive particles of all kinds whose origin is still a mystery. Most of them are protons, some of which can have extremely high energies; much higher than those achieved in the most powerful particle accelerators.

A shower of about 100 particles produced by a cosmic ray which found its way into a bubble chamber

When these highly energetic "cosmic rays" hit the atmosphere of the earth, they collide with the nuclei of the atmosphere's air molecules and produce a great variety of secondary particles which either decay or undergo further collisions, thus creating more particles which collide and decay again, and so on, until the last of them reach the earth. In this way, a single proton plunging into the earth's atmosphere can give rise to a whole cascade of events in which its original kinetic

by accident. The roughly horizontal tracks in the picture belong to the particles coming out of the accelerator.

energy is transformed into a shower of various particles, and is gradually absorbed as they penetrate the air undergoing multiple collisions. The same phenomenon that can be observed in the collision experiments of high-energy physics thus occurs naturally but more intensely all the time in the earth's atmosphere; a continual flow of energy going through a great variety of particle patterns in a rhythmic dance of creation and destruction. On the preceding page is a magnificent picture of this energy dance which was taken by accident when an unexpected cosmic-ray shower hit a bubble chamber at the European research center CERN during an experiment.

The processes of creation and destruction occurring in the world of particles are not only those which can be seen in the bubble-chamber photographs. They also include the creation and destruction of virtual particles which are exchanged in particle interactions and do not live long enough to be observed. Take, for example, the creation of two pions in a collision between a proton and an antiproton. A space-time diagram of this event would look like this (remember that the direction of time in these diagrams is from the bottom to the top!):

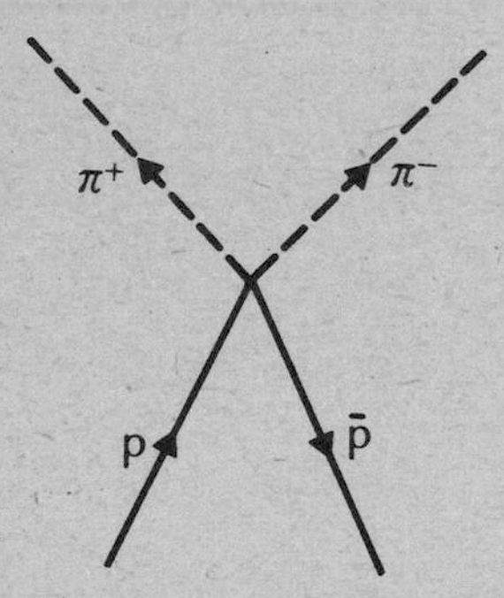

It shows the world lines of the proton (p) and the antiproton ($\bar{p}$) which collide at one point in space and time, annihilating each other and creating the two pions (π^+ and π^-). This diagram, however, does not give the full picture. The interaction between the proton and the antiproton can be pictured as the exchange of a virtual neutron, as the diagram shows opposite.

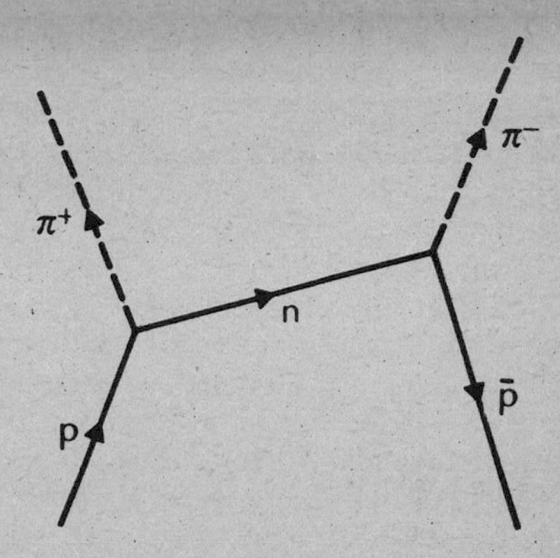

Similarly, the process shown in the following photograph, where four pions are created in a proton-antiproton

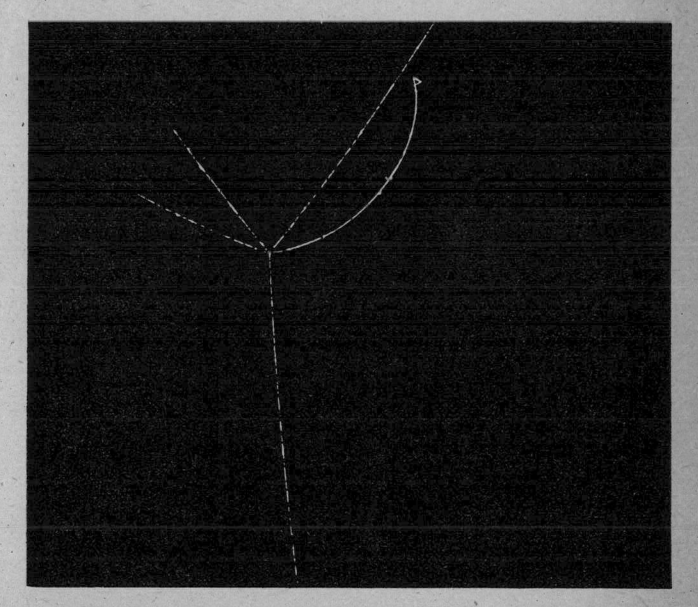

collision, can be pictured as a more complicated exchange process involving the creation and destruction of three virtual particles: two neutrons and one proton.

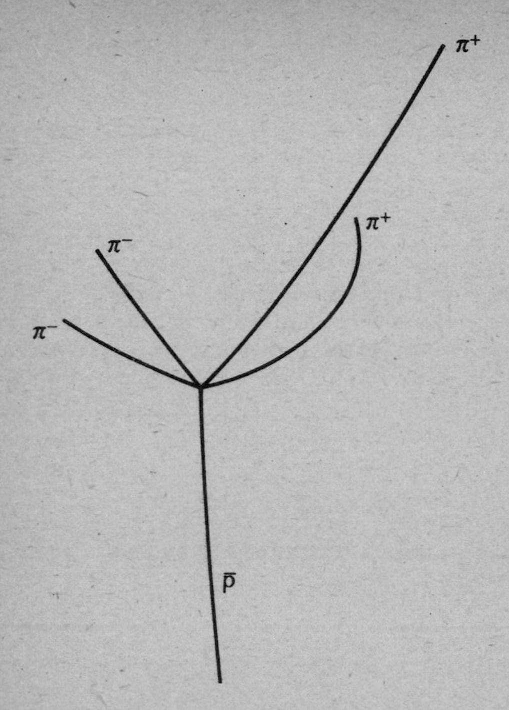

The corresponding Feynman diagram looks like this:*

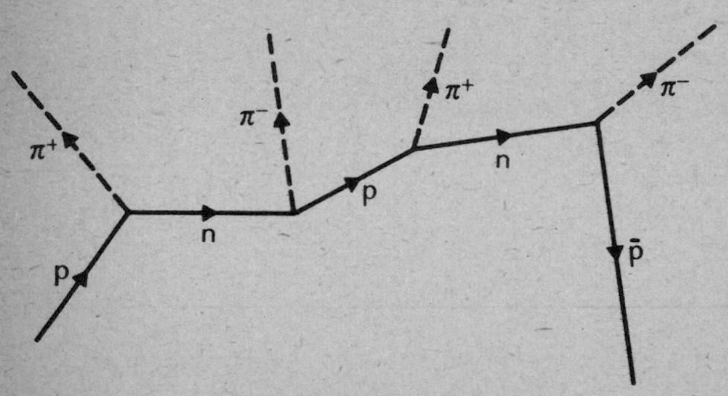

* The following diagrams are merely schematic and do not give the correct angles of the particle lines. Notice also that the initial proton sitting in the bubble chamber does not appear in the photograph, but has a world line in the space-time diagram because it moves in time.

These examples illustrate how the lines in the bubble-chamber photographs give only a rough picture of the particle interactions. The actual processes involve much more complicated networks of particle exchanges. The situation becomes, in fact, infinitely more complex when we remember that each of the particles involved in the interactions emits and reabsorbs virtual particles incessantly. A proton, for example, will emit and reabsorb a neutral pion every now and then; at other times, it may emit a π^+ and turn into a neutron which will absorb the π^+ after a short while and transform itself back into the proton. In the Feynman diagrams, the proton lines will in those cases have to be replaced by the following diagrams:

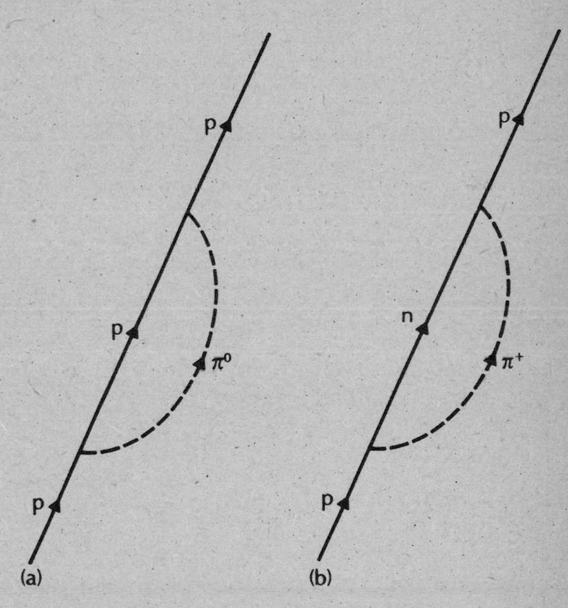

Feynman diagrams showing a proton emitting and reabsorbing virtual pions

In these virtual processes, the initial particle may disappear completely for a short time, as in diagram (b). A negative pion, to take another example, may create a neutron (n) plus an antiproton ($\bar{p}$) which then annihilate one another to reestablish the original pion:

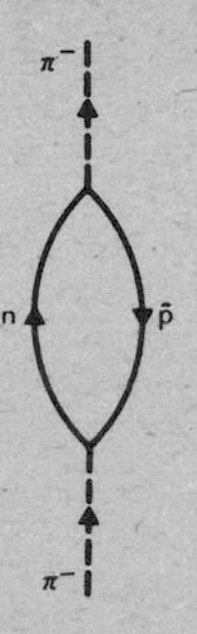

creation of a virtual neutron-antiproton pair

It is important to realize that all these processes follow the laws of quantum theory, and thus are tendencies, or probabilities, rather than actualities. Every proton exists potentially, i.e. with a certain probability, as a proton plus a π^0, as a neutron plus a π^+ and in many other ways. The examples shown above are only the simplest virtual processes. Much more complicated patterns arise when the virtual particles create other virtual particles, thus generating a whole network of virtual interactions.* In his book *The World of Elementary Particles*, Kenneth Ford has constructed a complicated example of such a network involving the creation and destruction of eleven virtual particles, and he comments on it: "[The diagram] pictures one such sequence of events, quite horrendous looking, but perfectly real. Every proton occasionally goes through exactly this dance of creation and destruction."[1]

Ford is not the only physicist to have used phrases like "dance of creation and destruction" and "energy dance." The ideas of rhythm and dance naturally come to mind when one tries to imagine the flow of energy going through the patterns that make up the particle world. Modern physics has shown us that movement and rhythm are essential properties of matter; that all matter,

* It should be noted that the possibilities are not completely arbitrary, but are restricted by several general laws to be discussed in the subsequent chapter.

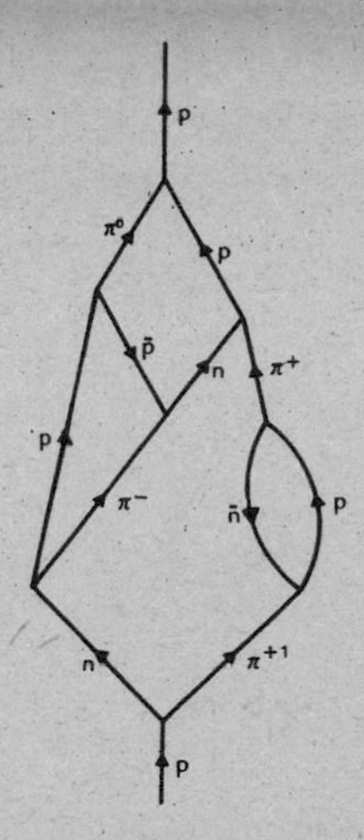

a network of virtual interactions; from Ford,
The World of Elementary Particles

whether here on earth or in outer space, is involved in a continual cosmic dance.

The Eastern mystics have a dynamic view of the universe similar to that of modern physics, and consequently it is not surprising that they, too, have used the image of the dance to convey their intuition of nature. A beautiful example of such an image of rhythm and dance is given by Alexandra David-Neel in her *Tibetan Journey*, where she describes how she met a lama who referred to himself as a "master of sound" and gave her the following account of his view of matter:

> All things . . . are aggregations of atoms that dance and by their movements produce sounds. When the rhythm of the dance changes, the sound it produces also changes. . . . Each atom perpetually sings its song, and the sound, at every moment, creates dense and subtle forms.[2]

The similarity of this view to that of modern physics becomes particularly striking when we remember that sound is a wave with a certain frequency which changes when the sound does, and that particles—the modern equivalent of the old concept of atoms—are also waves with frequencies proportional to their energies. According to field theory, each particle does indeed "perpetually sing its song," producing rhythmic patterns of energy (the virtual particles) in "dense and subtle forms."

The metaphor of the cosmic dance has found its most profound and beautiful expression in Hinduism in the image of the dancing god Shiva. Among his many incarnations, Shiva, one of the oldest and most popular Indian gods,* appears as the King of Dancers. According to Hindu belief, all life is part of a great rhythmic process of creation and destruction, of death and rebirth, and Shiva's dance symbolizes this eternal life-death rhythm which goes on in endless cycles. In the words of Ananda Coomaraswamy:

> In the night of *Brahman*, Nature is inert and cannot dance till Shiva wills it: He rises from His rapture, and dancing sends through inert matter pulsing waves of awakening sound, and lo! matter also dances, appearing as a glory round about Him. Dancing, He sustains its manifold phenomena. In the fullness of time, still dancing, He destroys all forms and names by fire and gives new rest. This is poetry, but nonetheless science.[3]

The Dance of Shiva symbolizes not only the cosmic cycles of creation and destruction, but also the daily rhythm of birth and death which is seen in Indian mysticism as the basis of all existence. At the same time, Shiva reminds us that the manifold forms in the world are *maya*—not fundamental, but illusory and ever-changing—as he keeps creating and dissolving them in the ceaseless flow of his dance. As Heinrich Zimmer has put it:

> His gestures wild and full of grace, precipitate the cosmic illusion; his flying arms and legs and the swaying of his torso produce—indeed, they are—the continuous creation-destruction of the universe, death exactly balancing birth, annihilation the end of every coming-forth.[4]

* See p. 80.

Shiva Nataraja, Brahmanical bronze,
South India, twelfth century.

Indian artists of the tenth and twelfth centuries have represented Shiva's cosmic dance in magnificent bronze sculptures of dancing figures with four arms whose superbly balanced and yet dynamic gestures express the rhythm and unity of Life. The various meanings of the dance are conveyed by the details of these figures in a complex pictorial allegory. The upper right hand of the god holds a drum to symbolize the primal sound of creation; the upper left bears a tongue of flame, the element of destruction. The balance of the two hands represents the dynamic balance of creation and destruction in the world, accentuated further by the Dancer's calm and detached face in the center of the two hands, in which the polarity of creation and destruction is dissolved and transcended. The second right hand is raised in the sign of "do not fear," symbolizing maintenance, protection and peace, while the remaining left hand points down to the uplifted foot which symbolizes release from the spell of *maya*. The god is pictured as dancing on the body of a demon, the symbol of man's ignorance which has to be conquered before liberation can be attained.

Shiva's dance—in the words of Coomaraswamy—is "the clearest image of the activity of God which any art or religion can boast of."[5] As the god is a personification of *Brahman*, his activity is that of *Brahman*'s myriad manifestations in the world. The dance of Shiva is the dancing universe; the ceaseless flow of energy going through an infinite variety of patterns that melt into one another.

Modern physics has shown that the rhythm of creation and destruction is not only manifest in the turn of the seasons and in the birth and death of all living creatures, but is also the very essence of inorganic matter. According to quantum field theory, all interactions between the constituents of matter take place through the emission and absorption of virtual particles. More than that, the dance of creation and destruction is the basis of the very existence of matter, since all material particles "self-interact" by emitting and reabsorbing virtual particles. Modern physics has thus revealed that every subatomic particle not only performs an energy dance, but also *is* an energy dance; a pulsating process of creation and destruction.

The patterns of this dance are an essential aspect of each particle's nature and determine many of its properties. For example, the energy involved in the emission and absorption of virtual particles is equivalent to a certain amount of mass which contributes to the mass of the self-interacting particle. Different particles develop different patterns in their dance, requiring different amounts of energy, and hence have different masses. Virtual particles, finally, are not only an essential part of all particle interactions and of most of the particles' properties, but are also created and destroyed by the vacuum. Thus, not only matter, but also the void, participates in the cosmic dance, creating and destroying energy patterns without end.

For the modern physicists, then, Shiva's dance is the dance of subatomic matter. As in Hindu mythology, it is a continual dance of creation and destruction involving the whole cosmos; the basis of all existence and of all natural phenomena. Hundreds of years ago, Indian artists created visual images of dancing Shivas in a beautiful series of bronzes. In our time, physicists have used the most advanced technology to portray the patterns of the cosmic dance. The bubble-chamber photographs of interacting particles, which bear testimony to the continual rhythm of creation and destruction in the universe, are visual images of the dance of Shiva equaling those of the Indian artists in beauty and profound significance. The metaphor of the cosmic dance thus unifies ancient mythology, religious art, and modern physics. It is indeed, as Coomaraswamy has said, "poetry, but nonetheless science."

16
QUARK SYMMETRIES
A NEW KOAN?

The subatomic world is one of rhythm, movement, and continual change. It is not, however, arbitrary and chaotic, but follows very definite and clear patterns. To begin with, all particles of a given kind are completely identical; they have exactly the same mass, electric charge, and other characteristic properties. Furthermore, all charged particles carry electric charges exactly equal (or opposite) to that of the electron, or charges of exactly twice that amount. The same is true of other quantities that are characteristic attributes of the particles; they do not take arbitrary values but are restricted to a limited number, which allows us to arrange the particles into a few distinct groups, or "families." This leads to the question of how these definite patterns arise in the dynamic and ever-changing particle world.

The emergence of clear patterns in the structure of matter is not a new phenomenon, but was already observed in the world of atoms. Like subatomic particles, atoms of a given kind are completely identical and the different kinds of atoms, of chemical elements, have been arranged into several groups in the periodic table. This classification is now well understood; it is based on the number of protons and neutrons present in the atomic nuclei and on the distribution of the electrons in spherical orbits, or "shells," around the nuclei. As discussed previously,* the wave nature of the electrons restricts the mutual distance of their orbits and the amount of rotation an electron can have in a given orbit to a few definite values corresponding to specific vibrations of the electron waves. Consequently, definite patterns arise in the atomic structure which are characterized by a set of integral "quantum numbers" and reflect the vibration patterns of the electron waves in their atomic orbits. These vibrations determine the "quantum states" of an atom and ascertain that any two atoms will be completely identical when they are both in their "ground state" or in the same "excited state."

The patterns in the particle world show great similarities to those in the world of atoms. Most particles, for example, spin about an axis like a top. Their spins are restricted to definite values which are integral multiples of some basic unit. Thus the baryons can only have spins of 1/2, 3/2, 5/2, etc., whereas the mesons have spins of 0, 1, 2, etc. This is strongly reminiscent of the amounts of rotation electrons are known to display in their atomic orbits, which are also restricted to definite values specified by integral numbers.

The analogy to the atomic patterns is further enforced by the fact that all strongly interacting particles, or hadrons, seem to fall into sequences whose members have identical properties except for their masses and spins. The higher members of these sequences are the extremely short-lived particles called "resonances" which have been discovered in great numbers over the past decade. The masses and spins of the resonances increase in a well-defined way within each sequence, which seem to extend indefinitely. These regularities suggest an analogy to the excited states of atoms and have led physicists to see the higher members of a hadron sequence not as different particles, but merely as excited states of the member with the lowest mass. Like an atom, a hadron can thus exist in various short-lived excited states involving higher amounts of rotation (or spin) and energy (or mass).

The similarities between the quantum states of atoms

* See p. 59.

and hadrons suggest that hadrons, too, are composite objects with internal structures that are capable of being "excited," that is, of absorbing energy to form a variety of patterns. At present, however, we do not understand how these patterns are formed. In atomic physics, they can be explained in terms of the properties and mutual interactions of the atom's constituents (the protons, neutrons and electrons), but in particle physics such an explanation has not yet been possible. The patterns found in the particle world have been determined and classified in a purely empirical way and cannot yet be derived from the details of the particles' structure.

The essential difficulty particle physicists have to face lies in the fact that the classical notion of composite "objects" consisting of a definite set of "constituent parts" cannot be applied to subatomic particles. The only way to find out what the "constituents" of these particles are is to break them up by banging them together in collision processes involving high energies. When this is done, however, the resulting fragments are never "smaller pieces" of the original particles. Two protons, for example, can break up into a great variety of fragments when they collide with high velocities, but there will never be "fractions of a proton" among them. The fragments will always be entire hadrons which are formed out of the kinetic energies and masses of the colliding protons. The decomposition of a particle into its "constituents" is thus far from being definite, depending, as it does, on the energy involved in the collision process. We are dealing here with a crucially relativistic situation where dynamic energy patterns are dissolved and rearranged, and the static concepts of composite objects and constituent parts cannot be applied to these patterns. The "structure" of a subatomic particle can only be understood in a dynamic sense; in terms of processes and interactions.

The way in which particles break up into fragments in collision processes is determined by certain rules, and as the fragments are again particles of the same kind, these rules can also be used to describe the regularities which can be observed in the particle world. In the sixties, when most of the presently known particles were discovered and "families" of particles began to appear, most physicists—quite naturally—concentrated their efforts on mapping out the emerging regularities, rather than tackling the arduous problem of finding the dynamic causes of the particle patterns. And in doing so, they were very successful.

The notion of symmetry played an important role in this research. By generalizing the common concept of symmetry and giving it a more abstract meaning, physicists were able to develop it into a powerful tool which proved extremely useful in the classification of particles.

In everyday life, the most common case of symmetry is associated with mirror reflection; a figure is said to be symmetric when you can draw a line through it and thereby divide it into two parts which are exact mirror images of each other. Higher degrees of symmetry are exhibited by patterns which allow several lines of symmetry to be drawn, like the following pattern used in Buddhist symbolism:

Reflection, however, is not the only operation associated with symmetry. A figure is also said to be symmetric if it looks the same after it has been rotated through a certain angle. The shape of the Chinese *yin-yang* diagram, for example, is based on such a rotational symmetry.

In particle physics, symmetries are associated with many other operations besides reflections and rotations, and these can take place not only in ordinary space (and time), but also in abstract mathematical spaces. They are applied to particles, or groups of particles, and since the particles' properties are inseparably linked to their mutual interactions, the symmetries also apply to the interactions: i.e., to the processes in which the particles are involved. The reason that these symmetry operations are so useful lies in the fact that they are closely related to "conservation laws." Whenever a process in the particle world displays a certain symmetry, there is a measurable quantity which is "conserved": a quantity which remains constant during the process. These quantities provide elements of constancy in the complex dance of subatomic matter and are thus ideal to describe the particle interactions. Some quantities are conserved in all interactions, others only in some of them, so that each process is associated with a set of conserved quantities. Thus the symmetries in the particles' properties appear as conservation laws in their interactions. Physicists use the two concepts interchangeably, referring sometimes to the symmetry of a process, sometimes to the corresponding conservation law, whichever is more convenient in the particular case.

There are four basic conservation laws which seem to be observed in all processes, three of them being connected with simple symmetry operations in ordinary space and time. All particle interactions are symmetric with respect to displacements in space—they will look exactly the same whether they take place in London or in New York. They are also symmetric with respect to displacements in time, which means they will occur in the same way on a Monday or on a Wednesday. The first of these symmetries is connected with the conservation of momentum, the second with the conservation of energy. This means that the total momentum of all particles involved in an interaction, and their total energy (including all their masses), will be exactly the same before and after the interaction. The third basic symmetry is one with respect to orientation in space. In a particle collision, for example, it does not make any difference whether the colliding particles approach each other along an axis oriented north-south or east-west. As a consequence of this symmetry, the total amount of rotation involved in a process (which includes the spins of the individual particles) is always conserved. Finally, there is the conservation of electric charge. It is connected with a more complicated symmetry operation, but in its formulation as a conservation law it is very simple: the total charge carried by all particles involved in an interaction remains constant.

There are several more conservation laws which correspond to symmetry operations in abstract mathematical spaces, like the one connected with charge conservation. Some of them hold for all interactions, as far as we know, others only for some of them (e.g., for strong and electromagnetic interactions, but not for weak interactions). The corresponding conserved quantities can be seen as "abstract charges" carried by the particles. Since they always take integer values (±1, ±2, etc.), or "half-integer" values (±1/2, ±3/2, ±5/2, etc.), they are called quantum numbers, in analogy to the quantum numbers in atomic physics. Each particle, then, is characterized by a set of quantum numbers which, in addition to its mass, specify its properties completely.

Hadrons, for example, carry definite values of "isospin" and "hypercharge," two quantum numbers which are conserved in all strong interactions. If the eight mesons listed in the table in the previous chapter are arranged according to the values of these two quantum numbers, they are seen to fall into a neat hexagonal pattern known as the "meson octet." This arrangement exhibits a great deal of symmetry; for example, particles and antiparticles occupy opposite places in the hexagon, the two particles in the center being their own antiparticles.

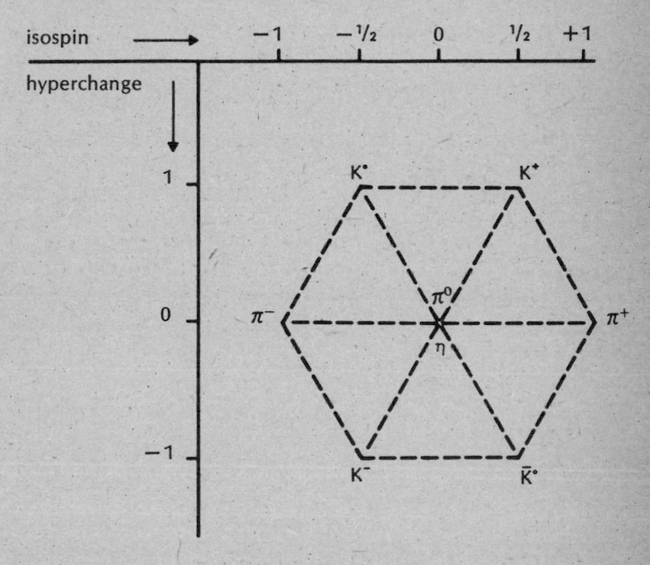

the meson octet

The eight lightest baryons form exactly the same pattern which is called the baryon octet. This time, however, the antiparticles are not contained in the octet, but form an identical "antioctet." The remaining baryon in our particle table—the omega—belongs to a different pattern, called the "baryon decuplet," together with nine resonances. All the particles in a given symmetry pattern have identical quantum numbers, except for isospin and hypercharge which give them their places in the pattern. For example, all mesons in the octet have zero spin (i.e., they do not spin at all); the baryons in the octet have a spin of 1/2, and those in the decuplet have a spin of 3/2.

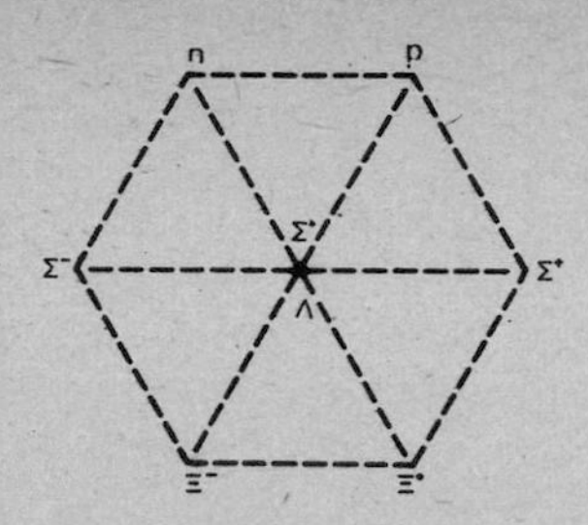

the baryon octet

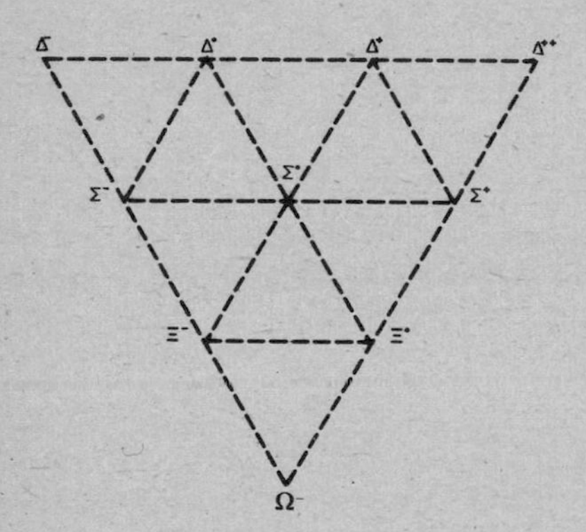

the baryon decuplet

The quantum numbers, then, are used to arrange particles into families forming neat symmetric patterns, to specify the places of the individual particles within each pattern, and at the same time to classify the various particle interactions according to the conservation laws they exhibit. The two related concepts of symmetry and conservation are thus seen to be extremely useful for expressing the regularities in the particle world.

It is surprising that most of these regularities can be represented in a very simple way if one assumes that all hadrons are made of a small number of elementary entities which have so far eluded direct observation. These entities have been given the fanciful name "quarks" by Murray Gell-Mann who referred his fellow physicists to the line in James Joyce's *Finnegans Wake*, "Three quarks for Muster Mark," when he postulated their existence. Gell-Mann succeeded in accounting for a large number of hadron patterns, such as the octets and the decuplet discussed above, by assigning appropriate quantum numbers to his three quarks and their antiquarks, and then putting these building blocks together in various combinations to form baryons and mesons whose quantum numbers are obtained simply by adding those of their constituent quarks. In this sense, baryons can be said to "consist of" three quarks, their antiparticles of the corresponding antiquarks, and mesons of a quark plus an antiquark.

The simplicity and efficiency of this model is striking, but it leads to severe difficulties if quarks are taken seriously as actual physical constituents of hadrons. So far, no hadrons have ever been broken up into their constituent quarks, in spite of bombarding them with the highest energies available, which means that quarks would have to be held together by extremely strong binding forces. According to our present understanding of particles and their interactions, these forces can only manifest themselves through the exchange of other particles, and consequently these other particles, too, would be present inside each hadron. If this were so, however, they would also contribute to the hadron's properties and thus destroy the simple additive scheme of the quark model.

In other words, if quarks are held together by strong interaction forces, these must involve other particles and the quarks must consequently show some kind of "structure," just like all the other strongly interacting particles. For the quark model, however, it is essential to have pointlike, structureless quarks. Because of this fundamental difficulty, it has so far not been possible to formulate the quark model in a consistent dynamic way which accounts for the symmetries and for the binding forces.

On the experimental side, there has been a fierce, but so far, unsuccessful "hunt for the quark" over the past decade. If single quarks exist, they should be quite conspicuous because Gell-Mann's model requires them to possess some very unusual properties, like electric charges of 1/3 and 2/3 of that of the electron, which do not appear anywhere in the particle world. So far, no particles with these properties have been observed in spite of the most intensive search. This persistent failure to detect them experimentally, plus the serious theoretical objections to their existence, have made the reality of quarks extremely doubtful.

On the other hand, the quark model continues to be very successful in accounting for the regularities found in the particle world, although it is no longer used in its original simple form. In Gell-Mann's original model, all hadrons could be built from three kinds of quarks and their antiquarks, but in the meantime physicists have had to postulate additional quarks to account for the great variety of hadron patterns. Gell-Mann himself recently proposed that each quark can appear in three different varieties which he called—most appropriately in a lecture in Paris—"red, white, and blue quarks." This increased the total number of quarks to nine, and since then three more quarks have been postulated,* which allowed one of the speakers at a recent physics conference to refer to them facetiously as "the twelve observed quarks."

The great number of regularities that can be successfully described in terms of these twelve quarks is truly impressive. There can be no doubt that hadrons exhibit "quark symmetries," even though our present understanding of particles and interactions precludes the existence of physical quarks. At present, in the summer of 1974, the paradoxes surrounding the quark model are becoming increasingly sharp. A great deal of experimental data support the quark model; others contradict it violently. No one has ever seen a quark, and according to our basic ideas about particle interactions quarks cannot exist. Yet, hadrons very often behave exactly as if they consisted of pointlike elementary constituents. This situation is strongly reminiscent of the early days of atomic physics when equally striking paradoxes led the physicists to a major breakthrough in their understanding of atoms. The quark puzzle has all the traits of a new *koan* which,

* The fourth triplet of quarks implies the existence of a new type of hadrons. The recently discovered "psi particles" may be of that kind.

in turn, could lead to a major breakthrough in our understanding of subatomic particles.

The discovery of symmetric patterns in the particle world has led many physicists to believe that these patterns reflect the fundamental laws of nature. During the past fifteen years, a great deal of effort has been devoted to the search for an ultimate "fundamental symmetry" that would incorporate all known particles and thus "explain" the structure of matter. This aim reflects a philosophical attitude which has been inherited from the ancient Greeks and cultivated throughout many centuries. Symmetry, together with geometry, played an important role in Greek science, philosophy and art, where it was identified with beauty, harmony and perfection. Thus the Pythagoreans regarded symmetric number patterns as the essence of all things; Plato believed that the atoms of the four elements had the shapes of regular solids, and most Greek astronomers thought that the heavenly bodies moved in circles because the circle was the geometrical figure with the highest degree of symmetry.

The attitude of Eastern philosophy with regard to symmetry is in striking contrast to that of the ancient Greeks. Mystical traditions in the Far East frequently use symmetric patterns as symbols or as meditation devices, but the concept of symmetry does not seem to play any major role in their philosophy. Like geometry, it is thought to be a construct of the mind, rather than a property of nature, and thus of no fundamental importance. Accordingly, many Eastern art forms have a striking predilection for asymmetry and often avoid all regular or geometrical shapes. The Zen-inspired paintings of China and Japan, often executed in the so-called "one-corner" style, or the irregular arrangements of flagstones in Japanese gardens clearly illustrate this aspect of Far Eastern culture.

It would seem, then, that the search for fundamental symmetries in particle physics is part of our Hellenistic heritage which is, somehow, inconsistent with the general world-view that begins to emerge from modern science. The emphasis on symmetry, however, is not the only aspect of particle physics. In contrast to the "static" symmetry approach, there has always been a "dynamic"

Flagstones in Katsura Palace Grounds, Kyoto, Japan.

"Birds by the Lake," by Liang K'ai, Southern Sung dynasty.

school of thought which does not regard the particle patterns as fundamental features of nature, but attempts to understand them as a consequence of the dynamic nature and essential interrelation of the subatomic world. The remaining two chapters show how this school of thought has given rise, in the past decade, to a radically different view of symmetries and laws of nature which is in harmony with the world view of modern physics described so far and which is in perfect agreement with Eastern philosophy.

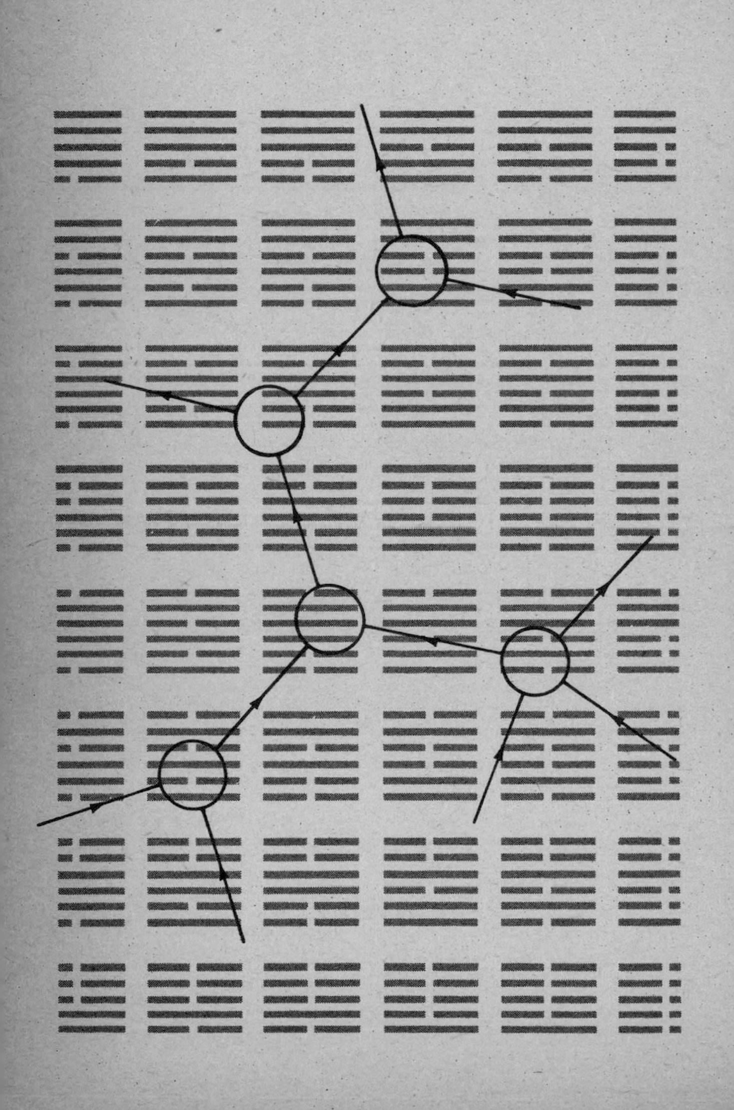

17
PATTERNS OF CHANGE

To explain the symmetries in the particle world in terms of a dynamic model—that is, one describing the interactions between the particles—is one of the major challenges of present-day physics. The problem, ultimately, is how to take into account simultaneously quantum theory and relativity theory. The particle patterns seem to reflect the "quantum nature" of the particles, since similar patterns occur in the world of atoms. In particle physics, however, they cannot be explained as wave patterns in the framework of quantum theory, because the energies involved are so high that relativity theory has to be applied. Only a "quantum-relativistic" theory of particles, therefore, can be expected to account for the observed symmetries.

Quantum field theory was the first model of that kind. It gave an excellent description of the electromagnetic interactions between electrons and photons, but it is much less appropriate for the description of strongly interacting particles. As more and more of these particles were discovered, physicists soon realized that it was highly unsatisfactory to associate each of them with a fundamental field, and when the particle world revealed itself as an increasingly complex tissue of interconnected processes, they had to look for other models to represent this dynamic and ever-changing reality. What was needed was a mathematical formalism which would be able to describe in a dynamic way the great variety of hadron patterns: their continual transformation into one another, their mutual interaction through the exchange of other particles, the formation of "bound states" of two or more hadrons, and their decay into various particle combinations. All these processes, which are often given the general name "particle reactions," are essential features of the strong interactions and have to be accounted for in a quantum-relativistic model of hadrons.

The framework which seems to be most appropriate for the description of hadrons and their interactions is called "S-matrix theory." Its key concept, the "S matrix," was originally proposed by Heisenberg in 1943 and has been developed, over the past two decades, into a complex mathematical structure which seems to be ideally suited to describe the strong interactions. The S matrix is a collection of probabilities for all possible reactions involving hadrons. It derives its name from the fact that one can imagine the whole assemblage of possible hadron reactions arranged in an infinite array of the kind mathematicians call a matrix. The letter S is a remainder of the original name, "scattering matrix" which refers to collision—or "scattering"—processes, the majority of particle reactions.

In practice, of course, one is never interested in the entire collection of hadron processes, but always in a few specific reactions. Therefore, one never deals with the whole S matrix, but only with those of its parts, or "elements," which refer to the processes under consideration. These are represented symbolically by diagrams like the one below which pictures one of the simplest and

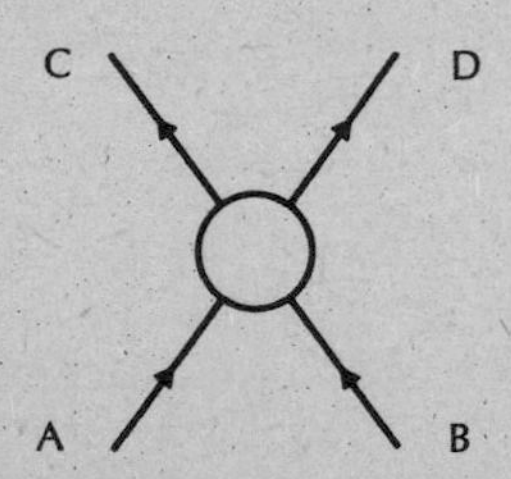

most general particle reactions: two particles, A and B, undergo a collision to emerge as two different particles, C and D. More complicated processes involve a greater number of particles and are represented by diagrams like the following:

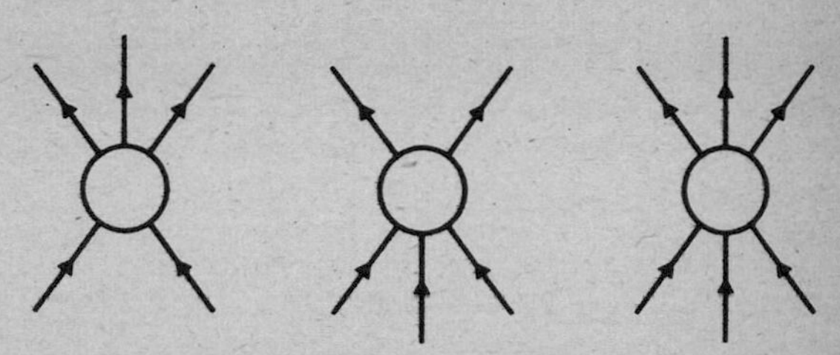

It must be emphasized that these S-matrix diagrams are very different from the Feynman diagrams of field theory. They do not picture the detailed mechanism of the reaction, but merely specify the initial and final particles. The standard process A + B → C + D, for example, might be pictured in field theory as the exchange of a virtual particle V, whereas in S-matrix theory, one simply draws

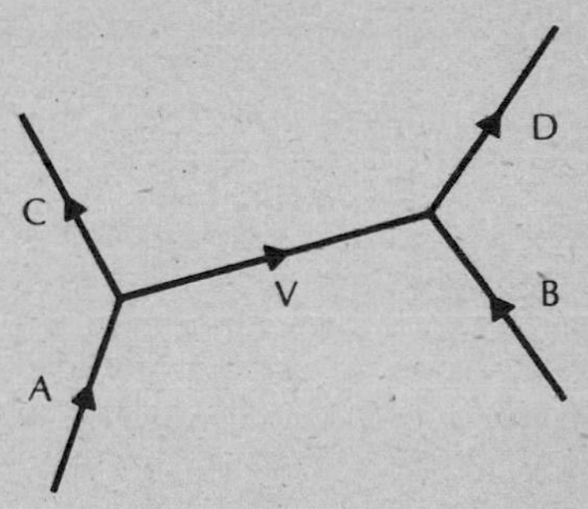

a circle without specifying what goes on inside it. Furthermore, the S-matrix diagrams are not space-time diagrams, but more general symbolic representations of particle reactions. These reactions are not assumed to take place at definite points in space and time, but are described in terms of the velocities (or, more precisely, in terms of the momenta) of the incoming and outgoing particles.

This means, of course, that an S-matrix diagram contains much less information than a Feynman diagram. On the other hand, S-matrix theory avoids a difficulty which is characteristic of field theory. The combined effects of quantum and relativity theory make it impossible to localize an interaction between definite particles precisely. Due to the uncertainty principle, the uncertainty of a particle's velocity will increase as its region of interaction is localized more sharply,* and consequently the amount of its kinetic energy will be increasingly uncertain. Eventually this energy will become large enough for new particles to be created, in accordance with relativity theory, and then one can no longer be certain of dealing with the original reaction. Therefore, in a theory which combines both quantum and relativity theories, it is not possible to specify the position of individual particles precisely. If this is done, as in field theory, one has to put up with mathematical inconsistencies which are, indeed, the main problem in all quantum field theories. S-matrix theory bypasses this problem by specifying the momenta of the particles and remaining sufficiently vague about the region in which the reaction occurs.

The important new concept in S-matrix theory is the shift of emphasis from objects to events; its basic concern is not with the particles, but with their reactions. Such a shift from objects to events is required both by quantum theory and by relativity theory. On the one hand, quantum theory has made it clear that a subatomic particle can only be understood as a manifestation of the interaction between various processes of measurement. It is not an isolated object but rather an occurrence, or event, which interconnects other events in a particular way. In the words of Heisenberg:

> [In modern physics], one has now divided the world not into different groups of objects but into different groups of connections. . . . What can be distinguished is the kind of connection which is primarily important in a certain phenomenon. . . . The world thus appears as a complicated tissue of events, in which connections of different kinds alternate or overlap or combine and thereby determine the texture of the whole.[1]

* See p. 142.

Relativity theory, on the other hand, has forced us to conceive of particles in terms of space-time: as four-dimensional patterns, as processes rather than objects. The S-matrix approach combines both of these viewpoints. Using the four-dimensional mathematical formalism of relativity theory, it describes all properties of hadrons in terms of reactions (or, more precisely, in terms of reaction probabilities), and thus establishes an intimate link between particles and processes. Each reaction involves particles which link it to other reactions and thus build up a whole network of processes.

A neutron, for example, may participate in two successive reactions involving different particles; the first, say, a proton and a π^-, the second a Σ^- and a K^+. The neutron thus interconnects these two reactions and integrates them into a larger process (see diagram (a) below). Each of the initial and final particles in this

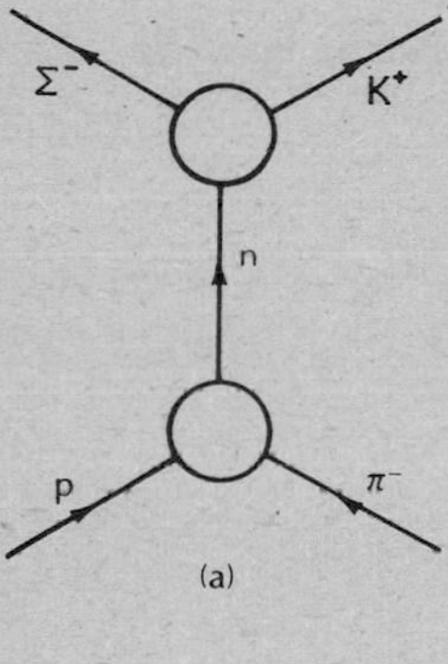

(a)

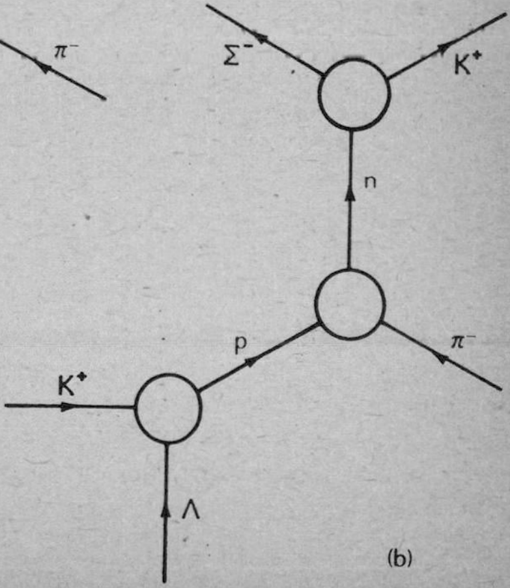

(b)

process will be involved in other reactions; the proton, for example, may emerge from an interaction between a K^+ and a Λ (see diagram (b) on preceding page); the K^+ in the original reaction may be linked to a K^- and a π^0; the π^- to three more pions.

The original neutron is thus seen to be part of a whole network of interactions; of a "tissue of events," all described by the S matrix. The interconnections in such a network cannot be determined with certainty, but are associated with probabilities. Each reaction occurs with some probability, which depends on the available energy and on the characteristics of the reaction, and these probabilities are given by the various elements of the S matrix.

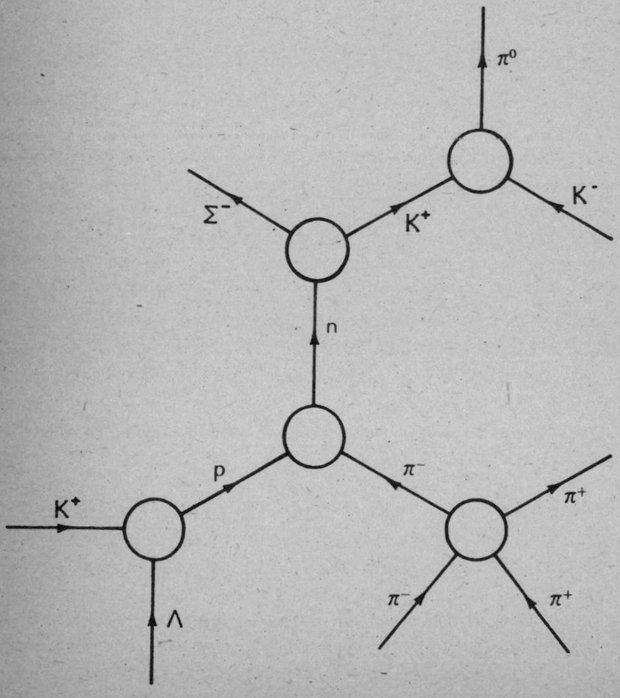

This approach allows one to define the structure of a hadron in a thoroughly dynamic way. The neutron in our network, for example, can be seen as a "bound state" of the proton and the π^- from which it arises, and also as a bound state of the Σ^- and the K^- into which it disintegrates. Either of these hadron combinations, and many others, may form a neutron, and consequently they can be said to be components of the neutron's "structure." The structure of a hadron, therefore, is not understood as a definite arrangement of constituent parts, but is given by all sets of particles which may interact with one another to form the hadron under consideration. Thus a proton exists potentially as a neutron-pion pair, a kaon-lambda pair, and so on. The proton also has the potential of disintegrating into any of these particle combinations if enough energy is available. The tendencies of a hadron to exist in various manifestations are expressed by the probabilities for the corresponding reactions, all of which may be regarded as aspects of the hadron's internal structure.

By defining the structure of a hadron as its tendency to undergo various reactions, S-matrix theory gives the concept of structure an essentially dynamic connotation. At the same time, this notion of structure is in perfect agreement with the experimental facts. Whenever hadrons are broken up in high-energy collision experiments, they disintegrate into combinations of other hadrons; thus they can be said to "consist" potentially of these hadron combinations. Each of the particles emerging from such a collision will, in turn, undergo various reactions, thus building up a whole network of events which can be photographed in the bubble chamber. The picture on page 256 and the ones in Chapter 15 are examples of such networks of reactions.

Although it is a matter of chance which network will arise in a particular experiment, each network is nevertheless structured according to definite rules. These rules are the conservation laws mentioned before; only those reactions can occur in which a well-defined set of quantum numbers is conserved. To begin with, the total energy has to remain constant in every reaction. This means that a certain combination of particles can emerge from a reaction only if the energy carried into the reaction is high enough to provide the required masses. Furthermore,

the emerging group of particles must collectively carry exactly the same quantum numbers that have been carried into the reaction by the initial particles. For example, a proton and a π^-, carrying a total electric charge of zero, may be dissolved in a collision and rearranged to emerge as a neutron plus a π^0, but they cannot emerge as a neutron and a π^+, as this pair would carry a total charge of +1.

The hadron reactions, then, represent a flow of energy in which particles are created and dissolved, but the energy can flow only through certain "channels" char-

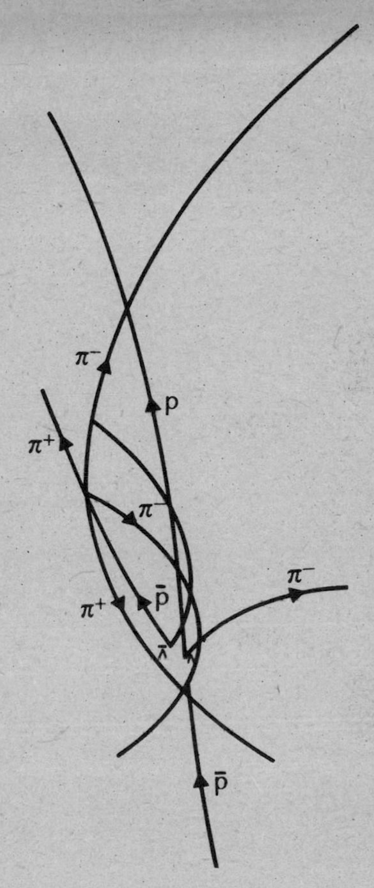

Opposite and above
a network of reactions involving protons, antiprotons, a lambda-antilambda pair, and several pions

acterized by the quantum numbers conserved in the strong interactions. In S-matrix theory, the concept of a reaction channel is more fundamental than that of a particle. It is defined as a set of quantum numbers which can be carried by various hadron combinations and often also by a single hadron. Which combination of hadrons flows through a particular channel is a matter of probability but depends, first of all, on the available energy. The diagram below, for example, shows an

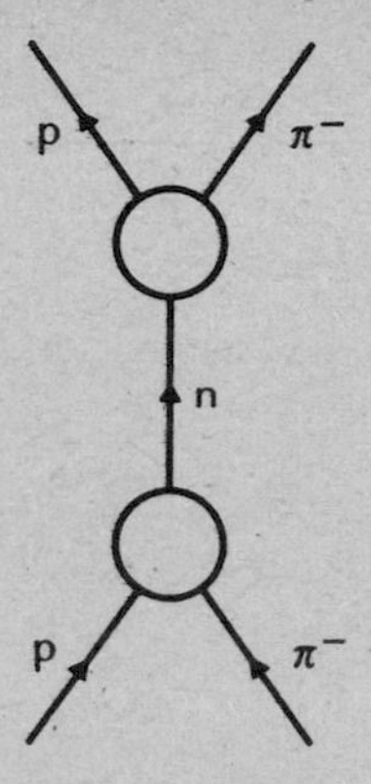

interaction between a proton and a π^- in which a neutron is formed as an intermediate state. Thus, the reaction channel is made up first by two hadrons, then by a single hadron, and finally by the initial hadron pair. The same channel can be made up, if more energy is available, by a Λ-K^0 pair, a Σ^--K^+ pair, and by various other combinations.

The notion of reaction channels is particularly appropriate to deal with resonances, those extremely short-lived hadron states which are characteristic of all strong interactions. They are such ephemeral phenomena that physicists were first reluctant to classify them as particles, and today the clarification of their properties still constitutes one of the major tasks in experimental high-energy physics. Resonances are formed in hadron collisions and disintegrate almost as soon as they come into being. They cannot be seen in the bubble chamber, but can be detected due to a very special behavior of reaction probabilities. The probability for two colliding hadrons to undergo a reaction—to interact with one another—depends on the energy involved in the collision. If the amount of this energy is modified, the probability will also change; it may increase or decrease with increasing energy, depending on the details of the reaction. At certain values of energy, however, the reaction probability is observed to increase sharply; a reaction is much more likely to occur at these values than at any other energy. This sharp increase is associated with the formation of a short-lived intermediate hadron with a mass corresponding to the energy at which the increase is observed.

The reason why these short-lived hadron states are called resonances is related to an analogy that can be drawn to the well-known resonance phenomenon encountered in connection with vibrations. In the case of sound, for example, the air in a cavity will in general respond only weakly to a sound wave coming from outside, but will begin to "resonate," or vibrate very strongly, when the sound wave reaches a certain frequency called the resonance frequency. The channel of a hadron reaction can be compared to such a resonant cavity, since the energy of the colliding hadrons is related to the frequency of the corresponding probability wave. When this energy, or frequency, reaches a certain value, the channel begins to resonate; the vibrations of the probability wave suddenly become very strong and thus cause a sharp increase in the reaction probability. Most reaction channels have several resonance energies, each of them corresponding to the mass of an ephemeral intermediate hadron state which is formed when the energy of the colliding particles reaches the resonance value.

In the framework of S-matrix theory, the problem of whether one should call the resonances "particles" or not does not exist. All particles are seen as intermediate states in a network of reactions, and the fact that the resonances live for a much shorter period than other hadrons does not make them fundamentally different. In fact, the word "resonance" is a very appropriate term. It applies both to the phenomenon in the reaction channel and to the hadron which is formed during that phenomenon, thus showing the intimate link between particles and reactions. A resonance is a particle, but not an object. It is much better described as an event, an occurrence, or a happening.

This description of hadrons in particle physics recalls to mind the words of D. T. Suzuki quoted above:*

* See p. 189.

"Buddhists have conceived an object as an event and not as a thing or substance." What Buddhists have realized through their mystical experience of nature has now been rediscovered through the experiments and mathematical theories of modern science.

In order to describe all hadrons as intermediate states in a network of reactions, one has to be able to account for the forces through which they mutually interact. These are the strong-interaction forces which deflect, or "scatter," colliding hadrons, dissolve and rearrange them in different patterns, and bind groups of them together to form intermediate bound states. In S-matrix theory, as in field theory, the interaction forces are associated with particles, but the concept of virtual particles is not used. Instead, the relation between forces and particles is based on a special property of the S matrix known as "crossing." To illustrate this property, consider the following diagram picturing the interaction between a proton and a π^-:

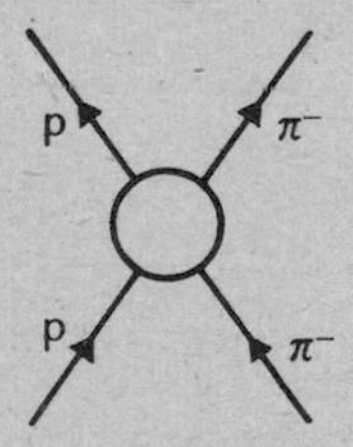

If this diagram is rotated through 90°, and if we keep the convention adopted previously,* that arrows pointing downward indicate antiparticles, the new diagram will

* See p. 168.

represent a reaction between an antiproton ($\bar{p}$) and a proton (p) which emerge from it as a pair of pions, the π^+ being the antiparticle of the π^- in the original reaction.

The "crossing" property of the S matrix now refers to the fact that both these processes are described by the same S-matrix element. This means that the two diagrams represent merely two different aspects, or "channels," of the same reaction.* Particle physicists are used to switching from one channel to the other in their calculations, and instead of rotating the diagrams they just read them upward or across from the left, and talk about the "direct channel" and the "cross channel." Thus the reaction in our example is read as $p+\pi^- \rightarrow p+\pi^-$ in the direct channel, and as $\bar{p}+p \rightarrow \pi^- + \pi^+$ in the cross channel.

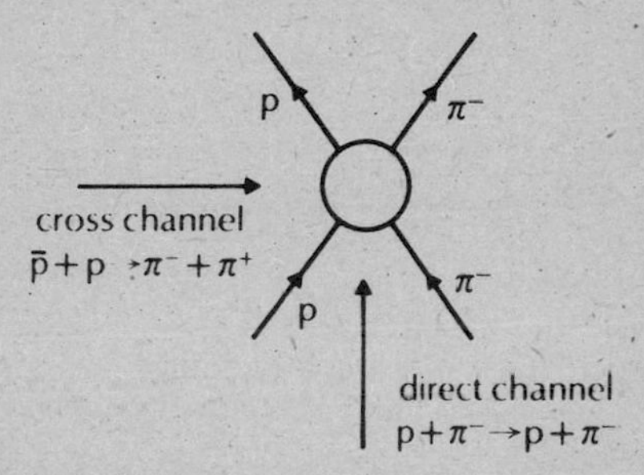

The connection between forces and particles is established through the intermediate states in the two channels. In the direct channel of our example, the proton and the π^- can form an intermediate neutron, whereas the cross channel can be made up by an intermediate neutral pion (π^0). This pion—the intermediate state in the cross channel—is interpreted as the manifestation of the force which acts in the direct channel binding the proton and the π^- together to form the neutron. Thus both channels are needed to associate the forces with particles; what appears as a force in one channel is manifest as an intermediate particle in the other.

* In fact, the diagram can be rotated further, and individual lines can be "crossed" to obtain different processes which are still described by the same S-matrix element. Each element represents altogether six different processes, but only the two mentioned above are relevant for our discussion of interaction forces.

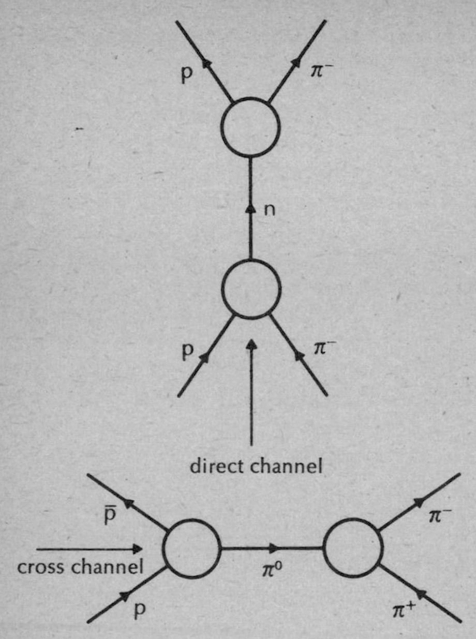

Although it is relatively easy to switch from one channel to the other mathematically, it is extremely difficult—if at all possible—to have an intuitive picture of the situation. This is because "crossing" is an essentially relativistic concept arising in the context of the four-dimensional formalism of relativity theory, and thus very difficult to visualize. A similar situation occurs in field theory where the interaction forces are pictured as the exchange of virtual particles. In fact, the diagram showing the intermediate pion in the cross channel is reminiscent of the Feynman diagrams picturing these particle exchanges,* and one might say, loosely speaking, that the proton and the π^- interact "through the exchange of a π^0." Such words are often used by physicists, but they do not fully describe the situation. An adequate description can only

* It should be remembered, however, that S-matrix diagrams are not space-time diagrams but symbolic representations of particle reactions. The switching from one channel to the other takes place in an abstract mathematical space.

be given in terms of direct and cross channels, that is, in abstract concepts which are almost impossible to visualize.

In spite of the different formalism, the general notion of an interaction force in S-matrix theory is quite similar to that in field theory. In both theories, the forces manifest themselves as particles whose mass determines the range of the force,* and in both theories they are recognized as intrinsic properties of the interacting particles; they reflect the structure of the particles' virtual clouds in field theory, and are generated by bound states of the interacting particles in S-matrix theory. The parallel to the Eastern view of forces discussed previously** applies thus to both theories. This view of interaction forces, furthermore, implies the important conclusion that all known particles must have some internal structure, because only then can they interact with the observer and thus be detected. In the words of Geoffrey Chew, one of the principal architects of S-matrix theory, "A truly elementary particle—completely devoid of internal structure—could not be subject to any forces that would allow us to detect its existence. The mere knowledge of a particle's existence, that is to say, implies that the particle possesses internal structure!"[2]

A particular advantage of the S-matrix formalism is the fact that it is able to describe the "exchange" of a whole family of hadrons. As mentioned in the previous chapter, all hadrons seem to fall into sequences whose members have identical properties except for their masses and spins. A formalism proposed originally by Tullio Regge makes it possible to treat each of these sequences as a single hadron existing in various excited states. In recent years, it has been possible to incorporate the Regge formalism into the S-matrix framework where it has been used very successfully for the description of hadron reactions. This has been one of the most important developments in S-matrix theory and can be seen as a first step toward a dynamic explanation of particle patterns.

* See p. 206.
** See p. 208.

The framework of the S matrix, then, is able to describe the structure of hadrons, the forces through which they mutually interact, and some of the patterns they form, in a thoroughly dynamic way in which each hadron is understood as an integral part of an inseparable network of reactions. The main challenge, and so far unsolved problem, in S-matrix theory is to use this dynamic description to account for the symmetries which give rise to the hadron patterns and conservation laws discussed in the previous chapter. In such a theory, the hadron symmetries would be reflected in the mathematical structure of the S matrix in such a way that it contains only elements which correspond to reactions allowed by the conservation laws. These laws would then no longer have the status of empirical regularities but would be a consequence of the S-matrix structure, and thus a consequence of the dynamic nature of hadrons.

At present, physicists are trying to achieve this ambitious aim by postulating several general principles which restrict the mathematical possibilities of constructing S-matrix elements and thus give the S matrix a definite structure. So far, three of these general principles have been established. The first is suggested by relativity theory and by our macroscopic experience of space and time. It says that the reaction probabilities (and thus the S-matrix elements) must be independent of displacements of the experimental apparatus in space and time, independent of its orientation in space, and independent of the state of motion of the observer. As discussed in the previous chapter, the independence of a particle reaction with regard to changes of orientation and displacements in space and time implies the conservation of the total amount of rotation, momentum and energy involved in the reaction. These "symmetries" are essential for our scientific work. If the results of an experiment changed according to where and when it was performed, science in its present form would be impossible. The last requirement, finally—that the experimental results must not depend on the observer's motion—is the principle of relativity which is the basis of relativity theory.*

* See p. 153.

The second general principle is suggested by quantum theory. It asserts that the outcome of a particular reaction can only be predicted in terms of probabilities and, furthermore, that the sum of the probabilities for all possible outcomes—including the case of no interaction between the particles—must be equal to one. In other words, we can be certain that the particles will either interact with one another, or not. This seemingly trivial statement turns out to be, in fact, a very powerful principle, known under the name of "unitarity," which severely restricts the possibilities of constructing S-matrix elements.

The third and final principle is related to our notions of cause and effect and is known as the principle of causality. It states that energy and momentum are transferred over spatial distances only by particles, and that this transfer occurs in such a way that a particle can be created in one reaction and destroyed in another only if the latter reaction occurs after the former. The mathematical formulation of the causality principle implies that the S matrix depends in a smooth way on the energies and momenta of the particles involved in a reaction, except for those values at which the creation of new particles becomes possible. At those values, the mathematical structure of the S matrix changes abruptly; it encounters what mathematicians call a "singularity." Each reaction channel contains several of these singularities; that is, there are several values of energy and momentum in each channel at which new particles can be created. The "resonance energies" mentioned before are examples of such values.

The fact that the S matrix exhibits singularities is a consequence of the causality principle, but the location of the singularities is not determined by it. The values of energy and momentum at which particles can be created are different for different reaction channels and depend on the masses and other properties of the created particles. The locations of the singularities thus reflect the properties of these particles, and since all hadrons can be created in particle reactions, the singularities of the S matrix mirror all the patterns and symmetries of hadrons.

The central aim of S-matrix theory is, therefore, to derive the singularity structure of the S matrix from the general principles. Up to now, it has not been possible to construct a mathematical model which satisfies all three principles, and it may well be that they are sufficient to determine all the properties of the S matrix—and thus all the properties of hadrons—uniquely.* If this turns out to be the case, the philosophical implications of such a theory would be very profound. All three of the general principles are related to our methods of observation and measurement, that is, to the scientific framework. If they are sufficient to determine the structure of hadrons, this would mean that the basic structures of the physical world are determined, ultimately, by the way in which we look at this world. Any fundamental change in our observational methods would imply a modification of the general principles which would lead to a different structure of the S matrix, and would thus imply a different structure of hadrons.

Such a theory of subatomic particles reflects the impossibility of separating the scientific observer from the observed phenomena, which has already been discussed in connection with quantum theory,** in its most extreme form. It implies, ultimately, that the structures and phenomena we observe in nature are nothing but creations of our measuring and categorizing mind.

That this is so is one of the fundamental tenets of Eastern philosophy. The Eastern mystics tell us again and again that all things and events we perceive are creations of the mind, arising from a particular state of consciousness and dissolving again if this state is transcended. Hinduism holds that all shapes and structures around us are created by a mind under the spell of *maya*, and it regards our tendency to attach deep significance to them as the basic human illusion. Buddhists call this illusion *avidya*, or ignorance, and see it as the state of a "defiled" mind. In the words of Ashvaghosha:

> When the oneness of the totality of things is not

* This conjecture, known as the "bootstrap" hypothesis, will be discussed in more detail in the subsequent chapter.
** See p. 126.

> recognized, then ignorance as well as particularization arises, and all phases of the defiled mind are thus developed. . . . All phenomena in the world are nothing but the illusory manifestation of the mind and have no reality on their own.[3]

This is also the recurring theme of the Buddhist Yogacara school which holds that all forms we perceive are "mind only"; projections, or "shadows" of the mind:

> Out of mind spring innumerable things, conditioned by discrimination. . . . These things people accept as an external world. . . . What appears to be external does not exist in reality; it is indeed mind that is seen as multiplicity; the body, property, and above—all these, I say, are nothing but mind.[4]

In particle physics, the derivation of the hadron patterns from the general principles of S-matrix theory is a long and arduous task, and so far only a few small steps have been taken toward achieving it. Furthermore, the theory in its present form cannot be applied to the electromagnetic interactions that give rise to the atomic structures and dominate the world of chemistry and biology. Nevertheless, the possibility that the hadron patterns will someday be derived from the general principles, and thus be seen to depend on our scientific framework, must be taken seriously. It is an exciting conjecture that this may be a general feature of particle physics which will also appear in future theories of electromagnetic, weak, and gravitational interactions. If this turns out to be true, modern physics will have come a long way toward agreeing with the Eastern sages that the structures of the physical world are *maya*, or 'mind only.'

S-matrix theory comes very close to Eastern thought not only in its ultimate conclusion, but also in its general view of matter. It describes the world of subatomic particles as a dynamic network of events and emphasizes change and transformation rather than fundamental structures or entities. In the East, such an emphasis is particularly strong in Buddhist thought where all things are seen as dynamic, impermanent and illusory. Thus S. Radhakrishnan writes:

> How do we come to think of things, rather than of processes in this absolute flux? By shutting our eyes to the successive events. It is an artificial attitude that makes sections in the stream of change, and calls them things. . . . When we shall know the truth of things, we shall realize how absurd it is for us to worship isolated products of the incessant series of transformations as though they were eternal and real. Life is no thing or state of a thing, but a continuous movement or change.[5]

Both the modern physicist and the Eastern mystic have realized that all phenomena in this world of change and transformation are dynamically interrelated. Hindus and Buddhists see this interrelation as a cosmic law, the law of *karma*, but they are generally not concerned with any specific patterns in the universal network of events. Chinese philosophy, on the other hand, which also emphasizes movement and change, has developed the notion of dynamic patterns which are continually formed and dissolved again in the cosmic flow of the *Tao*. In the *I Ching*, or *Book of Changes*, these patterns have been elaborated into a system of archetypal symbols, the so-called hexagrams.

The basic ordering principle of the patterns in the *I Ching** is the interplay of the polar opposites *yin* and *yang*. The *yang* is represented by a solid line (———), the yin by a broken line (— —), and the whole system of hexagrams is built up naturally from these two lines. By combining them in pairs, four configurations are obtained,

and by adding a third line to each of these, eight "trigrams" are generated:

* See p. 98.

In ancient China, the trigrams were considered to represent all possible cosmic and human situations. They were given names reflecting their basic characteristics—such as "The Creative," "The Receptive," "The Arousing," etc.—and they were associated with many images taken from nature and from social life. They represented, for example, heaven, earth, thunder, water, etc., as well as a family consisting of father, mother, three sons, and three daughters. They were, furthermore, associated with the cardinal points and with the seasons of the year, and were often arranged as follows:

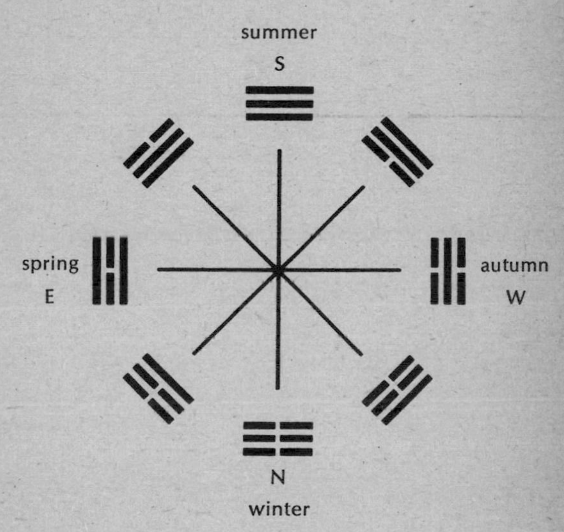

In this arrangement, the eight trigrams are grouped around a circle in the "natural order" in which they were generated, starting from the top (where the Chinese always place the south) and placing the first four trigrams on the left side of the circle, the second four on the right side. This arrangement shows a high degree of symmetry, opposite trigrams having *yin* and *yang* lines interchanged.

In order to increase the number of possible combinations further, the eight trigrams were combined in pairs by placing one above the other. In this way, sixty-four hexagrams were obtained, each consisting of six solid or broken lines. The hexagrams were arranged in several

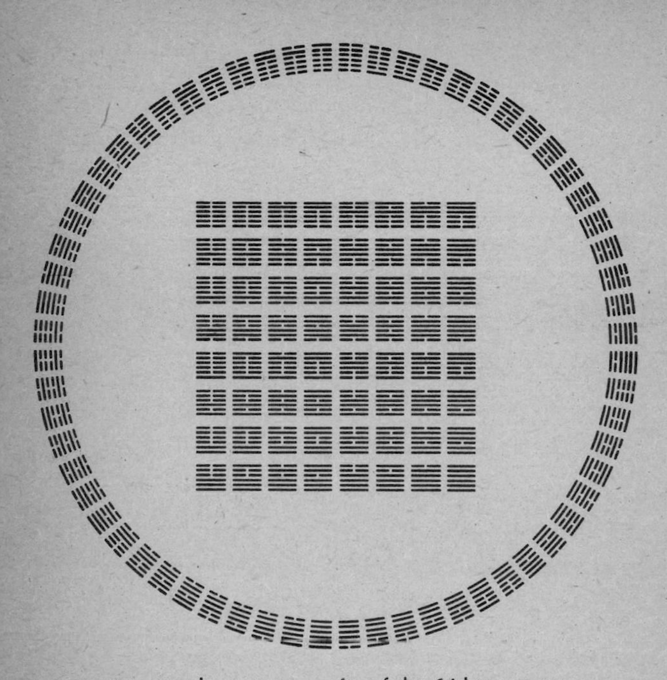

two regular arrangements of the 64 hexagrams

regular patterns, among which the two illustrated above were the most common; a square of eight times eight hexagrams, and a circular sequence showing the same symmetry as the circular arrangement of the trigrams.

The sixty-four hexagrams are the cosmic archetypes on which the use of the *I Ching* as an oracle book is based. For the interpretation of any hexagram, the various meanings of its two trigrams have to be taken into account. For example, when the trigram "The Arousing" is situated above the trigram "The Receptive," the hexagram is

the Arousing

the Receptive

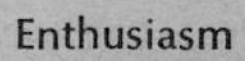

Enthusiasm

interpreted as movement meeting with devotion and thus inspiring enthusiasm, which is the name given to it.

To give another example, the hexagram for Progress represents "The Clinging" above "The Receptive," which is interpreted as the sun rising over the earth and thus as a symbol of rapid, easy progress.

In the *I Ching*, the trigrams and hexagrams represent the patterns of the *Tao* which are generated by the dynamic interplay of the *yin* and the *yang*, and are reflected in all cosmic and human situations. These situations, therefore, are not seen as static, but rather as stages in a continuous flow and change. This is the basic idea of the *Book of Changes* which is expressed in its very title. All things and situations in the world are subject to change and transformation, and so are their images, the trigrams and hexagrams. They are in a state of continual transition; one changing into another, solid lines pushing outward and breaking in two broken lines pushing inward and growing together.

Because of its notion of dynamic patterns, generated by change and transformation, the *I Ching* is perhaps the closest analogy to S-matrix theory in Eastern thought. In both systems, the emphasis is on processes rather than objects. In S-matrix theory, these processes are the particle reactions that give rise to all the phenomena in the world of hadrons. In the *I Ching*, the basic processes are called "the changes" and are seen as essential for an understanding of all natural phenomena:

> The changes are what have enabled the holy sages to reach all depths and to grasp the seeds of all things.[6]

These changes are not regarded as fundamental laws imposed on the physical world, but rather—in the words of Hellmut Wilhelm—as "an inner tendency according to which development takes place naturally and spontaneously."[7] The same can be said of the "changes" in the particle world. They, too, reflect the inner tendencies of the particles which are expressed, in S-matrix theory, in terms of reaction probabilities.

The changes in the world of hadrons give rise to structures and symmetric patterns which are represented symbolically by the reaction channels. Neither the structures nor the symmetries are regarded as fundamental features of the hadron world, but are seen as consequences of the particles' dynamic nature; that is, of their tendencies for change and transformation.

In the *I Ching*, too, the changes give rise to structures—the trigrams and hexagrams. Like the channels of particle reactions, these are symbolic representations of patterns of change. As the energy flows through the reaction channels, the "changes" flow through the lines of the hexagrams:

> Alteration, movement without rest,
> Flowing through the six empty places,
> Rising and sinking without fixed law,
> . . . .
> It is only change that is at work here.[8]

In the Chinese view, all things and phenomena around us arise out of the patterns of change and are represented by the various lines of the trigrams and hexagrams. Thus the things in the physical world are not seen as static, independent objects, but merely as transitional stages in the cosmic process which is the *Tao*:

> The *Tao* has changes and movements. Therefore the lines are called changing lines. The lines have gradations; therefore they represent things.[9]

As in the world of particles, the structures generated by the changes can be arranged in various symmetric patterns, such as the octagonal pattern formed by the eight trigrams, in which opposite trigrams have *yin* and *yang* lines interchanged. This pattern is even vaguely similar to the meson octet discussed in the previous chapter, in which particles and antiparticles occupy opposite places. The important point, however, is not this accidental similarity, but the fact that both modern physics and ancient Chinese thought consider change and transformation as the *primary* aspect of nature, and see the structures and symmetries generated by the changes as secondary. As he explains in the introduction to his translation of the *I Ching*, Richard Wilhelm regards this idea as the fundamental concept of the *Book of Changes*:

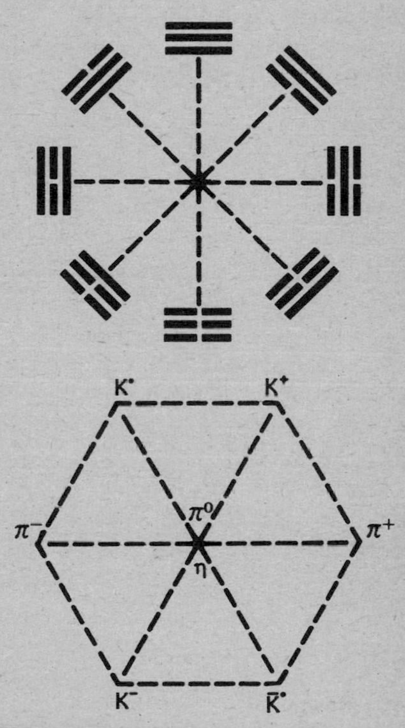

> The eight trigrams . . . were held to be in a state of continual transition, one changing into another, just as transition from one phenomenon to another is continually taking place in the physical world. Here we have the fundamental concept of the Book of Changes. The eight trigrams are symbols standing for changing transitional states; they are images that

> are constantly undergoing change. Attention centers not on things in their state of being—as is chiefly the case in the Occident—but upon their movements in change. The eight trigrams therefore are not representations of things as such but of their tendencies in movement.[10]

In modern physics, we have come to see the "things" of the subatomic world in very much the same way, laying stress upon movement, change and transformation and regarding the particles as transient stages in an ongoing cosmic process.

18
INTERPENETRATION

So far, our exploration of the world-view suggested by modern physics has repeatedly shown that the idea of "basic building blocks" of matter is no longer tenable. In the past, this concept was extremely successful in explaining the physical world in terms of a few atoms; the structures of the atoms in terms of a few nuclei surrounded by electrons; and finally, the structures of the nuclei in terms of two nuclear "building blocks," the proton and the neutron. Thus atoms, nuclei, and hadrons were, in turn, considered to be "elementary particles." None of them, however, fulfilled that expectation. Each time, these particles turned out to be composite structures themselves, and physicists hoped that the next generation of constituents would finally reveal themselves as the ultimate components of matter.

On the other hand, the theories of atomic and subatomic physics made the existence of elementary particles increasingly unlikely. They revealed a basic interconnection of matter, showing that energy of motion can be transformed into mass, and suggesting that particles are processes rather than objects. All these developments strongly indicated that the simple mechanistic picture of basic building blocks had to be abandoned, and yet many physicists are still reluctant to do so. The age-old tradition of explaining complex structures by breaking them down into simpler constituents is so deeply ingrained in Western thought that the search for these basic components is still going on.

There is, however, a radically different school of thought in particle physics which starts from the idea that nature cannot be reduced to fundamental entities, such as elementary particles or fundamental fields. It has to be understood entirely through its self-consistency, with its components being consistent both with one another and with themselves. This idea has arisen in the context of S-matrix theory and is known as the "bootstrap" hypothesis. Its originator and main advocate is Geoffrey Chew who, on the one hand, has developed the idea into a general "bootstrap" philosophy of nature and, on the other, has used it (in collaboration with other physicists) to construct specific models of particles formulated in S-matrix language. Chew has described the bootstrap hypothesis in several articles[1] which provide the basis for the following presentation.

The bootstrap philosophy constitutes the final rejection of the mechanistic world view in modern physics. Newton's universe was constructed from a set of basic entities with certain fundamental properties, which had been created by God and thus were not amenable to further analysis. In one way or another, this notion was implicit in all theories of natural science until the bootstrap hypothesis stated explicitly that the world cannot be understood as an assemblage of entities which cannot be analyzed further. In the new world-view, the universe is seen as a dynamic web of interrelated events. None of the properties of any part of this web is fundamental; they all follow from the properties of the other parts, and the overall consistency of their mutual interrelations determines the structure of the entire web.

Thus, the bootstrap philosophy represents the culmination of a view of nature that arose in quantum theory with the realization of an essential and universal interrelationship, acquired its dynamic content in relativity theory, and was formulated in terms of reaction probabilities in S-matrix theory. At the same time, this view of nature came ever closer to the Eastern world view and is now in harmony with Eastern thought, both in its general philosophy and in its specific picture of matter.

The bootstrap hypothesis not only denies the existence of fundamental constituents of matter, but accepts no fundamental entities whatsoever—no fundamental laws, equations, or principles—and thus abandons another idea which has been an essential part of natural science for hundreds of years. The notion of fundamental laws of nature was derived from the belief in a divine lawgiver which was deeply rooted in the Judaeo-Christian tradition. In the words of Thomas Aquinas:

> There is a certain Eternal Law, to wit, Reason, existing in the mind of God and governing the whole universe.[2]

This notion of an eternal, divine law of nature greatly influenced Western philosophy and science. Descartes wrote about the "laws which God has put into nature," and Newton believed that the highest aim of his scientific work was to give evidence of the "laws impressed upon nature by God." To discover the ultimate fundamental laws of nature remained the aim of natural scientists for the three centuries following Newton.

In modern physics, a very different attitude has now developed. Physicists have come to see that all their theories of natural phenomena, including the "laws" they describe, are creations of the human mind; properties of our conceptual map of reality, rather than of reality itself. This conceptual scheme is necessarily limited and approximate,* as are all the scientific theories and "laws of nature" it contains. All natural phenomena are ultimately interconnected, and in order to explain any one of them we need to understand all the others, which is obviously impossible. What makes science so successful is the discovery that approximations are possible. If one is satisfied with an approximate "understanding" of nature, one can describe selected groups of phenomena in this way, neglecting other phenomena which are less relevant. Thus one can explain many phenomena in terms of a few, and consequently understand different aspects of nature in an approximate way without having to understand everything at once. This is the scientific method; all scientific theories and models are approximations to the true nature of things, but the error involved in the approximation is often small enough to make such an approach meaningful. In particle physics, for example, the gravitational interaction forces between particles are usually ignored, as they are many orders of

* See pp. 15, 27.

magnitude weaker than those of the other interactions. Although the error caused by this omission is exceedingly small, it is clear that the gravitational interactions will have to be included in future, more accurate theories of particles.

Thus physicists construct a sequence of partial and approximate theories, each of them being more accurate than the previous one, but none of them representing a complete and final account of natural phenomena. Like these theories, all the "laws of nature" they describe are mutable, destined to be replaced by more accurate laws when the theories are improved. The incomplete character of a theory is usually reflected in its arbitrary parameters or "fundamental constants"; that is, in quantities whose numerical values are not explained by the theory, but have to be inserted into it after they have been determined empirically. Quantum theory cannot explain the value used for the mass of the electron, or field theory the magnitude of the electron's charge, or relativity theory that of the speed of light. In the classical view, these quantities were regarded as fundamental constants of nature which did not require any further explanation. In the modern view, their role of "fundamental constants" is seen as temporary and reflecting the limitations of the present theories. According to the bootstrap philosophy, they should be explained, one by one, in future theories as the accuracy and scope of these theories increase. Thus the ideal situation should be approached, but may never be reached, where the theory does not contain any unexplained "fundamental" constants, and where all its "laws" follow from the requirement of overall self-consistency.

It is important to realize, however, that even such an ideal theory must possess some unexplained features, although not necessarily in the form of numerical constants. As long as it is a scientific theory, it will require the acceptance, without explanation, of certain concepts which form the scientific language. To push the bootstrap idea further would lead beyond science:

> In the broad sense, the bootstrap idea, although fascinating and useful, is unscientific. . . . Science, as we know it, requires a language based on some

> unquestioned framework. Semantically, therefore, an attempt to explain *all* concepts can hardly be called "scientific."[3]

It is evident that the complete "bootstrap" view of nature, in which all phenomena in the universe are uniquely determined by mutual self-consistency, comes very close to the Eastern world-view. An indivisible universe, in which all things and events are interrelated, would hardly make sense unless it were self-consistent. In a way, the requirement of self-consistency, which forms the basis of the bootstrap hypothesis, and the unity and interrelation of all phenomena, which are so strongly emphasized in Eastern mysticism, are just different aspects of the same idea. This close connection is most clearly expressed in Taoism. For the Taoist sages, all phenomena in the world were part of the cosmic Way—the *Tao*—and the laws followed by the *Tao* were not laid down by any divine lawgiver, but were inherent in its nature. Thus we read in the *Tao Te Ching*:

> Man follows the laws of earth;
> Earth follows the laws of heaven;
> Heaven follows the laws of *Tao*;
> *Tao* follows the laws of its intrinsic nature.[4]

Joseph Needham, in his thorough study of Chinese science and civilization, discusses at great length how the Western concept of fundamental laws of nature, with its original implication of a divine lawgiver, has no counterpart in Chinese thought. "In the Chinese world-view," Needham writes, "the harmonious cooperation of all beings arose, not from the orders of a superior authority external to themselves, but from the fact that they were all parts in a hierarchy of wholes forming a cosmic pattern, and what they obeyed were the internal dictates of their own natures."[5]

According to Needham, the Chinese did not even have a word corresponding to the classical Western idea of a "law of nature." The term which comes closest to it is *li*, which the Neo-Confucian philosopher Chu Hsi*

* See p. 92.

describes as "the innumerable veinlike patterns included in the *Tao*."[6] Needham translates *li* as "principle of organization" and gives the following comments:

> In its most ancient meaning, it signified the pattern in things, the markings of jade or fibers in muscle. . . . It acquired the common dictionary meaning "principle," but always conserved the undertone of "pattern." . . . There is "law" implicit in it, but this law is the law to which parts of wholes have to conform by virtue of their very existence as parts of wholes. . . . The most important thing about parts is that they have to fit precisely into place with the other parts in the whole organism which they compose.[7]

It is easy to see how such a view led the Chinese thinkers to the idea which has only recently been developed in modern physics, that self-consistency is the essence of all laws of nature. The following passage by Ch'en Shun, an immediate pupil of Chu Hsi who lived around A.D. 1200, gives a very clear account of this idea in words which could be taken as a perfect explanation of the notion of self-consistency in the bootstrap philosophy:

> *Li* is a natural and unescapable law of affairs and things. . . . The meaning of "natural and unescapable" is that [human] affairs and [natural] things are made just exactly to fit into place. The meaning of "law" is that the fitting into place occurs without the slightest excess or deficiency. . . . The men of old, investigating things to the utmost, and searching out *li*, wanted to elucidate the natural inescapableness of [human] affairs and [natural] things, and this simply means that what they were looking for was all the exact places where things precisely fit together. Just that.[8]

In the Eastern view then, as in the view of modern physics, everything in the universe is connected to everything else and no part of it is fundamental. The properties of any part are determined, not by some fundamental law, but by the properties of all the other parts. Both physicists and mystics realize the resulting impossibility of fully explaining any phenomenon, but then they take different attitudes. Physicists, as discussed before, are satisfied with an approximate understanding of nature. The Eastern mystics, on the other hand, are not interested in approximate, or 'relative' knowledge. They are concerned with 'absolute' knowledge involving an understanding of the totality of Life. Being well aware of the essential interrelationship of the universe, they realize that to explain something means, ultimately, to show how it is connected to everything else. As this is impossible, the Eastern mystics insist that no single phenomenon can be explained. Thus Ashvaghosha:

> All things in their fundamental nature are not namable or explicable. They cannot be adequately expressed in any form of language.[9]

The Eastern sages, therefore, are generally not interested in explaining things, but rather in obtaining a direct nonintellectual experience of the unity of all things. This was the attitude of the Buddha, who answered all questions about life's meaning, the origin of the world, or the nature of *nirvana*, with a "noble silence." The nonsensical answers of Zen masters, when asked to explain something, seem to have the same purpose: to make the student realize that everything is a consequence of all the rest; that "explaining" nature just means to show its unity; that, ultimately, there is nothing to explain. When a monk asked Tozan, who was weighing some flax, "What is Buddha?" Tozan said, "This flax weighs three pounds";[10] and when Joshu was asked why Bodhidharma came to China, he replied, "An oak tree in the garden."[11]

To free the human mind from words and explanations is one of the main aims of Eastern mysticism. Both Buddhists and Taoists speak of a 'network of words,' or a 'net of concepts,' thus extending the idea of the interconnected web to the realm of the intellect. As long as we try to explain things, we are bound by *karma*: trapped in our conceptual network. To transcend words and explanations means to break the bonds of *karma* and attain liberation.

The world-view of the Eastern mystics shares with the bootstrap philosophy of modern physics not only an emphasis on the mutual interrelation and self-consistency of all phenomena, but also the denial of fundamental constituents of matter. In a universe which is an inseparable whole and where all forms are fluid and ever-changing, there is no room for any fixed fundamental entity. The notion of "basic building blocks" of matter is therefore generally not encountered in Eastern thought. Atomic theories of matter have never been developed in Chinese thought, and although they have risen in some schools of Indian philosophy, they are rather peripheral to Indian mysticism. In Hinduism, the notion of atoms is prominent in the Jaina system (which is regarded as unorthodox since it does not accept the authority of the Vedas). In Buddhist philosophy, atomic theories have arisen in two schools of Hinayana Buddhism, but are treated as illusory products of *avidya* by the more important Mahayana branch. Thus Ashvaghosha states:

> When we divide some gross [or composite] matter, we can reduce it to atoms. But as the atom will also be subject to further division, all forms of material existence, whether gross or fine, are nothing but the shadow of particularization and we cannot ascribe any degree of [absolute or independent] reality to them.[12]

The principal schools of Eastern mysticism thus agree with the view of the bootstrap philosophy that the universe is an interconnected whole in which no part is any more fundamental than the other, so that the properties of any one part are determined by those of all the others. In that sense, one might say that every part "contains" all the others and, indeed, a vision of mutual embodiment seems to be characteristic of the mystical experience of nature. In the words of Sri Aurobindo:

> Nothing to the supramental sense is really finite; it is founded on a feeling of all in each and of each in all.[13]

This notion of 'all in each and each in all' has found its most extensive elaboration in the *Avatamsaka* school of Mahayana Buddhism* which is often considered to be the final culmination of Buddhist thought. It is based on the *Avatamsaka Sutra*, traditionally believed to have been delivered by the Buddha while he was in deep meditation after his Awakening. This voluminous *sutra*, which has so far not been translated into any Western language, describes in great detail how the world is perceived in the enlightened state of consciousness, when "the solid outlines of individuality melt away and the feeling of finiteness no longer oppresses us."[14] In its last part, called the *Gandavyuha*, it tells the story of a young pilgrim, Sudhana, and gives the most vivid account of his mystical experience of the universe, which appears to him as a perfect network of mutual relations, where all things and events interact with each other in such a way that each of them contains, in itself, all the others. The following passage from the *sutra*, paraphrased by D. T. Suzuki, uses the image of a magnificently decorated tower to convey Sudhana's experience:

> The Tower is as wide and spacious as the sky itself. The ground is paved with [innumerable] precious stones of all kinds, and there are within the Tower [innumerable] palaces, porches, windows, staircases, railings, and passages, all of which are made of the seven kinds of precious gems. . . .
>
> And within this Tower, spacious and exquisitely ornamented, there are also hundreds of thousands . . . of towers, each one of which is as exquisitely ornamented as the main Tower itself and as spacious as the sky. And all these towers, beyond calculation in number, stand not at all in one another's way; each preserves its individual existence in perfect harmony with all the rest; there is nothing here that bars one tower being fused with all the others individually and collectively; there is a state of perfect intermingling and yet of perfect orderliness. Sudhana, the young pilgrim, sees himself in all the towers as well as in each single tower, where all is contained in one and each contains all.[15]

* See p. 89.

The Tower in this passage is, of course, a metaphor for the universe itself, and the perfect mutual interfusion of its parts is known in Mahayana Buddhism as 'interpenetration.' The *Avatamsaka* makes it clear that this interpenetration is an essentially dynamic interrelation which takes place not only spatially but also temporally. As mentioned previously,* space and time are also seen as interpenetrating.

The experience of interpenetration in the state of enlightenment can be seen as a mystical vision of the complete "bootstrap" situation, where all phenomena in the universe are harmoniously interrelated. In such a state of consciousness, the realm of the intellect is transcended and causal explanations become unnecessary, being replaced by the direct experience of the mutual interdependence of all things and events. The Buddhist concept of interpenetration thus goes far beyond any scientific bootstrap theory. Nevertheless, there are models of subatomic particles in modern physics, based on the bootstrap hypothesis, which show the most striking parallels to the views of Mahayana Buddhism.

When the bootstrap idea is formulated in a scientific context, it has to be limited and approximate, and its main approximation consists in neglecting all but the strong interactions. Since these interaction forces are about a hundred times stronger than the electromagnetic ones, and many more orders of magnitude stronger than weak and gravitational interactions, such an approximation seems reasonable. The scientific bootstrap, then, deals exclusively with strongly interacting particles, or hadrons, and is therefore often called the "hadron bootstrap." It is formulated in the framework of S-matrix theory, and its aim is to derive all properties of hadrons and their interactions uniquely from the requirement of self-consistency. The only "fundamental laws" accepted are the general S-matrix principles discussed in the previous chapter, which are required by our methods of observation and measurement and thus constitute the unquestioned framework necessary for all science. Other

* See p. 158.

properties of the S matrix may have to be postulated temporarily as "fundamental principles," but will be expected to emerge as a necessary consequence of self-consistency in the complete theory. The postulate that all hadrons form sequences described by the Regge formalism* may be of that kind.

In the language of S-matrix theory, then, the bootstrap hypothesis suggests that the full S matrix—and thus all the properties of hadrons—can be determined uniquely from the general principles because there is only one possible S matrix consistent with all three of them. This conjecture receives support from the fact that physicists have never come close to constructing a mathematical model which satisfies the three general principles. If the only consistent S matrix is the one describing *all* properties and interactions of hadrons, as the bootstrap hypothesis assumes, the physicists' failure to construct a consistent partial S matrix becomes understandable.

The phenomena involving hadrons are so complex that it is by no means certain whether the complete self-consistent S matrix will ever be constructed, but one can envisage a series of partially successful models of smaller scope. Each of them would be intended to cover only a part of hadron physics and would therefore contain some unexplained parameters representing its limitations, but the parameters of one model may be explained by another. Thus more and more hadron phenomena may gradually be covered with ever-increasing accuracy by a mosaic of interlocking models whose net number of unexplained parameters will keep decreasing. The adjective "bootstrap" is thus never appropriate for any individual model, but can be applied only to a combination of mutually consistent models, none of which is any more fundamental than the others. As Chew has put it, "A physicist who is able to view any number of different partially successful models without favoritism is automatically a bootstrapper."[16]

A number of partial models of that kind already exist, formulated in S-matrix language and describing certain aspects of hadron phenomena. The most successful of

* See p. 263.

them are the so-called "dual models" which make extensive use of the dual description of hadron reactions in terms of direct and cross channels.* These models incorporate, for the first time, two of the three general principles** plus the crossing property of the S matrix and the Regge formalism. They constitute so far the most promising attempts at carrying out the bootstrap program.

The picture of hadrons which emerges from these bootstrap models is often summed up in the provocative phrase, "Every particle consists of all other particles." It must not be imagined, however, that each hadron contains all the others in a classical, static sense. Rather than "containing" one another, hadrons "involve" one another in the dynamic and probabilistic sense of S-matrix theory, each hadron being a potential "bound state" of all sets of particles which may interact with one another to form the hadron under consideration.*** In that sense, all hadrons are composite structures whose components are again hadrons, and none of them is any more elementary than the others. The binding forces holding the structures together manifest themselves through the exchange of particles, and these exchanged particles are again hadrons. Each hadron, therefore, plays three roles: it is a composite structure, it may be a constituent of another hadron, and it may be exchanged between constituents and thus constitute part of the forces holding a structure together. The concept of "crossing" is crucial for this picture. Each hadron is held together by forces associated with the exchange of other hadrons in the cross channel, each of which is, in turn, held together by forces to which the first hadron makes a contribution. Thus, "each particle helps to generate other particles, which in turn generate it."[17] The whole set of hadrons generates itself in this way or pulls itself up, so to say, by its "bootstraps." The idea, then, is that this extremely complex bootstrap mechanism is self-determining: that is, that there is only one way in which it can be

* See p. 261.
** They do not satisfy the so-called "unitarity principle"; see p. 264.
*** See p. 254.

achieved. In other words, there is only one possible self-consistent set of hadrons—the one found in nature.

In the hadron bootstrap, all particles are dynamically composed of one another in a self-consistent way, and in that sense can be said to "contain" one another. In Mahayana Buddhism, a very similar notion is applied to the whole universe. This cosmic network of interpenetrating things and events is illustrated in the *Avatamsaka Sutra* by the metaphor of Indra's net, a vast network of precious gems hanging over the palace of the god Indra. In the words of Sir Charles Eliot:

> In the heaven of Indra, there is said to be a network of pearls, so arranged that if you look at one you see all the others reflected in it. In the same way each object in the world is not merely itself but involves every other object and in fact *is* everything else. "In every particle of dust, there are present Buddhas without number."[18]

The similarity of this image with that of the hadron bootstrap is indeed striking. The metaphor of Indra's net may justly be called the first bootstrap model, created by the Eastern sages some 2,500 years before the beginning of particle physics. Buddhists insist that the concept of interpenetration is not comprehensible intellectually, but is to be experienced by an enlightened mind in the state of meditation. Thus D. T. Suzuki writes:

> The Buddha [in the *Gandavyuha*] is no more the one who is living in the world conceivable in space and time. His consciousness is not that of an ordinary mind which must be regulated according to the senses and logic. . . . The Buddha of the *Gandavyuha* lives in a spiritual world which has its own rules.[19]

In modern physics, the situation is quite similar. The idea of every particle containing all the others is inconceivable in ordinary space and time. It describes a reality which, like the one of the Buddha, has its own rules. In the case of the hadron bootstrap, they are the rules of quantum theory and relativity theory, the key concept being that the forces holding particles together are themselves particles exchanged in the cross channels. This concept can be given a precise mathematical meaning, but is almost impossible to visualize. It is a specifically relativistic feature of the bootstrap, and since we have no direct experience of the four-dimensional world of space-time, it is extremely difficult to imagine how a single particle can contain all other particles and at the same time be part of each of them. This, however, is exactly the view of the Mahayana:

> When the one is set against all the others, the one is seen as pervading them all and at the same time embracing them all in itself.[20]

The idea of each particle containing all the others has not only arisen in Eastern mysticism, but also in Western mystical thought. It is implicit, for example, in William Blake's famous lines:

> To see a world in a grain of sand
> And a heaven in a wild flower,
> Hold infinity in the palm of your hand,
> And eternity in an hour.

Here again, a mystical vision has led to an image of the bootstrap type; if the poet sees the world in a grain of sand, the modern physicist sees the world in a hadron.

A similar image appears in the philosophy of Leibniz, who considered the world as being made of fundamental substances called "monads," each of which mirrored the whole universe. This led him to a view of matter which shows similarities to that of Mahayana Buddhism and to the hadron bootstrap.* In his *Monadology,* Leibniz writes:

> Each portion of matter may be conceived of as a garden full of plants, and as a pond full of fishes. But each branch of the plant, each member of the

* The parallels between Leibniz's view of matter and the hadron bootstrap have recently been discussed; see G. Gale, "Chew's Monadology," *Journal of History of Ideas,* Vol. 35 (April–June 1974), pp. 339–48.

animal, each drop of its humors, is also such a garden or such a pond.[21]

It is interesting that the similarity of these lines to the passages of the *Avatamsaka Sutra* mentioned before may stem from an actual Buddhist influence on Leibniz. Joseph Needham has argued[22] that Leibniz was well acquainted with Chinese thought and culture through translations he received from Jesuit monks, and that his philosophy might very well have been inspired by the Neo-Confucian school of Chu Hsi with which he was familiar. This school, however, has one of its roots in Mahayana Buddhism, and in particular in the *Avatamsaka* (Chinese: *Hua-yen*) school of the Mahayana branch. Needham, in fact, mentions the parable of Indra's net of pearls explicitly in connection with the Leibnizian monads.

A more detailed comparison of Leibniz's notion of "mirroring relations" between monads with the idea of interpenetration in the Mahayana seems to show, however, that the two are rather different, and that the Buddhist conception of matter comes much closer to the spirit of modern physics than that of Leibniz. The principal difference between the *Monadology* and the Buddhist view seems to be that the Leibnizian monads are fundamental substances which are seen as the ultimate constituents of matter. Leibniz begins the *Monadology* with the words, "The monad of which we shall here speak is merely a simple substance, which enters into composites; *simple*, that is to say, without parts." He goes on to say, "And these monads are the true atoms of nature, and, in a word, the elements of all things."[23] Such a "fundamentalist" view is in striking contrast to the bootstrap philosophy, and is also totally different from the view of Mahayana Buddhism which rejects all fundamental entities or substances. Leibniz's fundamentalist way of thinking is also reflected in his view of forces which he regards as laws "imprinted by divine decree" and essentially different from matter. "Forces and activity," he writes, "cannot be states of a merely passive thing like matter."[24] Again, this is contrary to the views of modern physics and of Eastern mysticism.

As far as the actual interrelation between the monads is concerned, the main difference to the hadron bootstrap seems to be that monads do not interact with each other; they "have no windows," as Leibniz says, and merely mirror one another. In the hadron bootstrap, on the other hand, as in the Mahayana, the emphasis is on the interaction, or "interpenetration," of all particles. Furthermore, the bootstrap and the Mahayana views of matter are both "space-time" views which see objects as events whose mutual interpenetration can only be understood if one realizes that space and time, too, are interpenetrating.

The bootstrap hypothesis is not yet firmly established and the technical difficulties involved in its implementation are considerable. Nevertheless, physicists already speculate about extending the self-consistent approach beyond the description of hadrons. In the present context of S-matrix theory, such an extension is not possible. The framework of the S matrix has been developed specifically to describe the strong interactions and cannot be applied to the rest of particle physics; the principal reason being that it cannot accommodate the massless particles which are characteristic of all the other interactions. To enlarge the hadron bootstrap, therefore, a more general framework will have to be found, and in this new framework some of the concepts which are at present accepted without explanation will have to be "bootstrapped"; they will have to be derived, that is, from the overall self-consistency. According to Geoffrey Chew, these might include our conception of macroscopic space-time and, perhaps, even that of human consciousness:

> Carried to its logical extreme, the bootstrap conjecture implies that the existence of consciousness, along with all other aspects of nature, is necessary for self-consistency of the whole.[25]

This view, again, is in perfect harmony with the views of the Eastern mystical traditions which have always regarded consciousness as an integral part of the universe. In the Eastern view, human beings, like all other life forms, are parts of an inseparable organic whole.

Their intelligence, therefore, implies that the whole, too, is intelligent. Man is seen as the living proof of cosmic intelligence; in us, the universe repeats over and over again its ability to produce forms through which it becomes consciously aware of itself.

In modern physics, the question of consciousness has arisen in connection with the observation of atomic phenomena. Quantum theory has made it clear that these phenomena can be understood only as links in a chain of processes, the end of which lies in the consciousness of the human observer.* In the words of Eugene Wigner, "It was not possible to formulate the laws of [quantum theory] in a fully consistent way without reference to consciousness."[26] The pragmatic formulation of quantum theory used by the scientists in their work does not refer to their consciousness explicitly. Wigner and other physicists have argued, however, that the explicit inclusion of human consciousness may be an essential aspect of future theories of matter.

Such a development would open exciting possibilities for a direct interaction between physics and Eastern mysticism. The understanding of one's consciousness and of its relation to the rest of the universe is the starting point of all mystical experience. The Eastern mystics have explored various modes of consciousness throughout centuries, and the conclusions they have reached are often radically different from the ideas held in the West. If physicists really want to include the nature of human consciousness in their realm of research, a study of Eastern ideas may well provide them with stimulating new viewpoints.

Thus the future enlargement of the hadron bootstrap, with the "bootstrapping" of space-time and of human consciousness it may require, opens up unprecedented possibilities which may well go beyond the conventional framework of science:

> Such a future step would be immensely more profound than anything comprising the hadron bootstrap; we would be obliged to confront the elusive concept of observation and, possibly, even that of

* See p. 126.

> consciousness. Our current struggle with the hadron bootstrap may thus be only a foretaste of a completely new form of human intellectual endeavor, one that will not only lie outside of physics but will not even be describable as "scientific."[27]

Where, then, does the bootstrap idea lead us? This, of course, nobody knows, but it is fascinating to speculate about its ultimate fate. One can imagine a network of future theories covering an ever-increasing range of natural phenomena with ever-increasing accuracy; a network which will contain fewer and fewer unexplained features, deriving more and more of its structure from the mutual consistency of its parts. Someday, then, a point will be reached where the only unexplained features of this network of theories will be the elements of the scientific framework. Beyond that point, the theory will no longer be able to express its results in words, or in rational concepts, and will thus go beyond science. Instead of a bootstrap *theory* of nature, it will become a bootstrap *vision* of nature, transcending the realms of thought and language; leading out of science and into the world of *acintya*, the unthinkable. The knowledge contained in such a vision will be complete, but cannot be communicated in words. It will be the knowledge which Lao Tzu had in mind, more than two thousand years ago, when he said:

> He who knows does not speak,
> He who speaks does not know.[28]

EPILOGUE

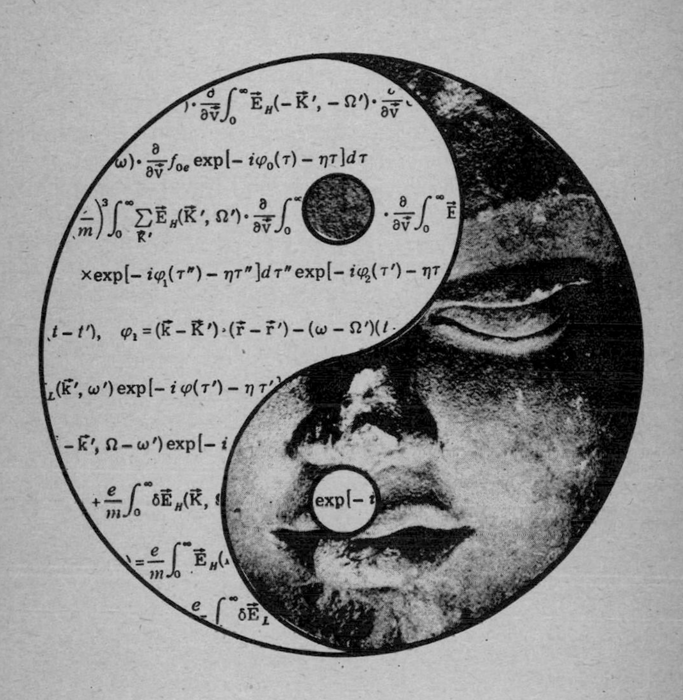

The Eastern religious philosophies are concerned with timeless mystical knowledge which lies beyond reasoning and cannot be adequately expressed in words. The relation of this knowledge to modern physics is but one of its many aspects and, like all the others, it cannot be demonstrated conclusively but has to be experienced in a direct intuitive way. What I hope to have achieved, to some extent, therefore, is not a rigorous demonstration, but rather to have given the reader an opportunity to relive, every now and then, an experience which has become for me a source of continuing joy and inspiration; that the principal theories and models of modern physics lead to a view of the world which is internally consistent and in perfect harmony with the views of Eastern mysticism.

For those who have experienced this harmony, the significance of the parallels between the world-views of physicists and mystics is beyond any doubt. The interesting question, then, is not *whether* these parallels exist, but *why*; and, furthermore, what their existence implies.

In trying to understand the mystery of Life, man has followed many different approaches. Among them, there are the ways of the scientist and mystic, but there are many more; the ways of poets, children, clowns, shamans, to name but a few. These ways have resulted in different descriptions of the world, both verbal and nonverbal, which emphasize different aspects. All are valid and useful in the context in which they arose. All of them, however, are only descriptions, or representations, of reality and are therefore limited. None can give a complete picture of the world.

The mechanistic world view of classical physics is useful for the description of the kind of physical phenomena we encounter in our everyday life and thus appropriate for dealing with our daily environment, and it has also proved extremely successful as a basis for technology. It is inadequate, however, for the description of physical phenomena in the submicroscopic realm. Opposed to the mechanistic conception of the world is the view of the mystics which may be epitomized by the word "organic," as it regards all phenomena in the universe as integral parts of an inseparable harmonious whole. This world-view emerges in the mystical traditions from meditative states of consciousness. In their description of the world, the mystics use concepts which are derived from these nonordinary experiences and are, in general, inappropriate for a scientific description of macroscopic phenomena. The organic world-view is not advantageous for constructing machines, nor for coping with the technical problems in an overpopulated world.

In everyday life, then, both the mechanistic and the organic views of the universe are valid and useful; the one for science and technology, the other for a balanced and fulfilled spiritual life. Beyond the dimensions of our everyday environment, however, the mechanistic concepts lose their validity and have to be replaced by organic concepts which are very similar to those used by the mystics. This is the essential experience of modern physics which has been the subject of our discussion. Physics in the twentieth century has shown that the concepts of the organic world view, although of little value for science and technology on the human scale, become extremely useful at the atomic and subatomic level. The organic view, therefore, seems to be more fundamental than the mechanistic. Classical physics, which is based on the latter, can be derived from quantum theory, which implies the former, whereas the reverse is not possible. This seems to give a first indication why we might expect the world views of modern physics and Eastern mysticism to be similar. Both emerge when man inquires into the essential nature of things—into the deeper realms of matter in physics; into the deeper realms of consciousness in mysticism—when he discovers a different reality behind the superficial mechanistic appearance of everyday life.

The parallels between the views of physicists and mystics become even more plausible when we recall the other similarities which exist in spite of their different approaches. To begin with, their method is thoroughly empirical. Physicists derive their knowledge from experiments; mystics from meditative insights. Both are observations, and in both fields these observations are acknowledged as the only source of knowledge. The object of observation is of course very different in the two cases. The mystic looks within and explores his or her consciousness at its various levels, which include the body as the physical manifestation of the mind. The experience of one's body is, in fact, emphasized in many Eastern traditions and is often seen as the key to the mystical experience of the world. When we are healthy, we do not feel any separate parts in our body but are aware of it as an integrated whole; and this awareness generates a feeling of well-being and happiness. In a similar way, the mystic is aware of the wholeness of the entire cosmos which is experienced as an extension of the body. In the words of Lama Govinda:

> To the enlightened man . . . whose consciousness embraces the universe, to him the universe becomes his "body," while his physical body becomes a manifestation of the Universal Mind, his inner vision an expression of the highest reality, and his speech an expression of eternal truth and mantric power.[1]

In contrast to the mystic, the physicist begins his inquiry into the essential nature of things by studying the material world. Penetrating into ever-deeper realms of matter, he has become aware of the essential unity of all things and events. More than that, he has also learnt that he himself and his consciousness are an integral part of this unity. Thus the mystic and the physicist arrive at the same conclusion; one starting from the inner realm, the other from the outer world. The harmony between their views confirms the ancient Indian wisdom that *Brahman*, the ultimate reality without, is identical to *Atman*, the reality within.

A further similarity between the ways of the physicist and mystic is the fact that their observations take place in realms which are inaccessible to the ordinary senses. In modern physics, these are the realms of the atomic and subatomic world; in mysticism they are nonordinary states of consciousness in which the sense world is transcended. Mystics often talk about experiencing higher dimensions in which impressions of different centers of consciousness are integrated into a harmonious whole. A similar situation exists in modern physics where a four-dimensional "space-time" formalism has been developed which unifies concepts and observations belonging to different categories in the ordinary three-dimensional world. In both fields, the multidimensional experiences transcend the sensory world and are therefore almost impossible to express in ordinary language.

We see that the ways of the modern physicist and the Eastern mystic, which seem at first totally unrelated, have, in fact, much in common. It should not be too surprising, therefore, that there are striking parallels in their descriptions of the world. Once these parallels between Western science and Eastern mysticism are accepted, a number of questions will arise concerning their implications. Is modern science, with all its sophisticated machinery, merely rediscovering ancient wisdom, known to the Eastern sages for thousands of years? Should physicists therefore abandon the scientific method and begin to meditate? Or can there be a mutual influence between science and mysticism—perhaps even a synthesis?

I think all these questions have to be answered in the negative. I see science and mysticism as two complementary manifestations of the human mind; of its rational and intuitive faculties. The modern physicist experiences the world through an extreme specialization of the rational mind; the mystic through an extreme specialization of the intuitive mind. The two approaches are entirely different and involve far more than a certain view of the physical world. However, they are complementary, as we have learned to say in physics. Neither is comprehended in the other, nor can either of them be reduced to the other; but both of them are necessary, supplementing one another for a fuller understanding of the world. To paraphrase an old Chinese saying, mystics understand the roots of the *Tao* but not its branches; scientists understand its branches but not its roots. Science does not need mysticism and mysticism does not need science, but man needs both. Mystical experience is necessary to understand the deepest nature of things, and science is essential for modern life. What we need, therefore, is not a synthesis, but a dynamic interplay between mystical intuition and scientific analysis.

So far, this has not been achieved in our society. At present, our attitude is too *yang*—to use again Chinese phraseology—too rational, male, and aggressive. Scientists themselves are a typical example. Although their theories are leading to a world-view which is similar to that of the mystics, it is striking how little this has affected the attitudes of most scientists. In mysticism, knowledge cannot be separated from a certain way of life which becomes its living manifestation. To acquire mystical knowledge means to undergo a transformation; one could even say that the knowledge *is* the transformation. Scientific knowledge, on the other hand, can often stay abstract and theoretical. Thus most of today's physicists do not seem to realize the philosophical, cultural, and spiritual implications of their theories. Many of them actively support a society which is still based on the mechanistic, fragmented world-view, without seeing that science points beyond such a view, toward a oneness of the universe which includes not only our natural environment but also our fellow human beings. I believe that the world-view implied by modern physics is inconsistent with our present society, which does not reflect the harmonious interrelatedness we observe in nature. To achieve such a state of dynamic balance, a radically different social and economic structure will be needed: a cultural revolution in the true sense of the word. The survival of our whole civilization may depend on whether we can bring about such a change. It will depend, ultimately, on our ability to adopt some of the *yin* attitudes of Eastern mysticism; to experience the wholeness of nature and the art of living with it in harmony.

NOTES

I THE WAY OF PHYSICS

Chapter 1: Modern Physics—A Path with a Heart?

1 J. R. Oppenheimer, *Science and the Common Understanding*, pp. 8–9.
2 N. Bohr, *Atomic Physics and Human Knowledge*, p. 20.
3 W. Heisenberg, *Physics and Philosophy*, p. 202.
4 Ashvaghosha, *The Awakening of Faith*, p. 78.
5 *Brihad-aranyaka Upanishad*, 3.7.15.

Chapter 2: Knowing and Seeing

1 W. Heisenberg, *Physics and Philosophy*, p. 125.
2 Chuang Tzu, trans. James Legge, ch. 26.
3 *Katha Upanishad*, 3.15.
4 *Kena Upanishad*, 3.
5 Quoted in J. Needham, *Science and Civilization in China*, vol. II, p. 85.
6 W. James, *The Varieties of Religious Experience*, p. 388.
7 B. Russell, *History of Western Philosophy*, p. 37.
8 D. T. Suzuki, *On Indian Mahayana Buddhism*, p. 237.
9 J. Needham, op. cit., vol. II, p. 33.
10 From the *Zenrin kushu*, in I. Muira & R. Fuller Sasaki, *The Zen Koan*, p. 103.
11 D. T. Suzuki, *Outlines of Mahayana Buddhism*, p. 235.
12 In Carlos Castaneda, *A Separate Reality*, p. 20.
13 Lao Tzu, *Tao Te Ching*, trans. Ch'u Ta-Kao, ch. 41.
14 Ibid., ch. 48.
15 Chuang Tzu, op. cit., ch. 13.
16 In P. Kapleau, *Three Pillars of Zen*, pp. 53–54.
17 A. K. Coomaraswamy, *Hinduism and Buddhism*, p. 33.
18 In A. W. Watts, *The Way of Zen*, p. 183.
19 Ibid., p. 187.

Chapter 3: Beyond Language

1 W. Heisenberg, *Physics and Philosophy*, p. 177.
2 D. T. Suzuki, *On Indian Mahayana Buddhism*, p. 239.
3 W. Heisenberg, op. cit., pp. 178–79.
4 In D. T. Suzuki, *The Essence of Buddhism*, p. 26.
5 In P. Kapleau, *Three Pillars of Zen*, p. 135.
6 W. Heisenberg, op. cit., p. 42.

Chapter 4: The New Physics

1 D. T. Suzuki, *The Essence of Buddhism*, p. 7.
2 W. Heisenberg, *Physics and Philosophy*, p. 167.
3 In P. A. Schilpp (ed.), *Albert Einstein: Philosopher-Scientist*, p. 45.
4 N. Bohr, *Atomic Physics and the Description of Nature*, p. 2.
5 S. Aurobindo, *On Yoga II*, Tome One, p. 327.
6 Quoted in M. Capek, *The Philosophical Impact of Contemporary Physics*, p. 7.
7 Ibid., p. 36.
8 In M. P. Crosland (ed.), *The Science of Matter*, p. 76.
9 Quoted in M. Capek, op. cit., p. 122.
10 Quoted in J. Jeans, *The Growth of Physical Science*, p. 237.
11 *Tables of Particle Properties*, published by the Particle Data Group in Physics Letters, Vol. 50B, No. 1, 1974.

II THE WAY OF EASTERN MYSTICISM

Chapter 5: Hinduism

1 *Mundaka Upanishad*, 2.2.3.
2 *Bhagavad Gita*, 4.42.
3 *Bhagavad Gita*, 13.12.
4 *Maitri Upanishad*, 6.17.
5 *Brihad-aranyaka Upanishad*, 1.4.6.
6 *Chandogya Upanishad*, 6.9.4.
7 *Bhagavad Gita*, 8.3.
8 Ibid., 3.27–28.
9 *Brihad-aranyaka Upanishad*, 4.3.21.

Chapter 6: Buddhism

1 *Dhammapada*, 113.
2 *Digha Nikaya*, ii, 154.
3 D. T. Suzuki, *On Indian Mahayana Buddhism*, p. 122.
4 D. T. Suzuki, *The Essence of Buddhism*, p. 54.

Chapter 7: Chinese Thought

1 Chuang Tzu, trans. James Legge, ch. 13.
2 J. Needham, *Science and Civilization in China*, vol. II, p. 35.

3 Fung Yu-Lan, *A Short History of Chinese Philosophy*, p. 14.
4 Chuang Tzu, op. cit., ch. 22.
5 Quoted in J. Needham, op. cit., vol. II, p. 51.
6 Lao Tzu, *Tao Te Ching*, trans. Ch'u Ta-Kao, chs. 40 and 25.
7 Ibid., ch. 29.
8 Wang Ch'ung, A.D. 80, quoted in J. Needham, op. cit., vol. IV, p. 7.
9 R. Wilhelm, *The I Ching or Book of Changes*, p. 297.
10 Kuei Ku Tzu, fourth century B.C., quoted in J. Needham, op. cit., vol. IV, p. 6.
11 Chuang Tzu, op. cit., ch. 22.
12 R. Wilhelm, op. cit., p. xlvii.
13 Ibid., p. 321.
14 Ibid., p. 348.

Chapter 8: Taoism

1 Chuang Tzu, trans. James Legge, ch. 22.
2 Ibid., ch. 24.
3 Ibid., ch. 2.
4 Ibid., ch. 13.
5 *Bhagavad Gita*, 2.45.
6 Quoted in Fung Yu-Lan, *A Short History of Chinese Philosophy*, p. 112.
7 Lao Tzu, *Tao Te Ching*, trans. Ch'u Ta-Kao, ch. 36.
8 Ibid., ch. 22.
9 Chuang Tzu, op. cit., ch. 17.
10 In G. S. Kirk, *Heraclitus—The Cosmic Fragments*, p. 307.
11 Ibid., pp. 105, 184.
12 Ibid., p. 149.
13 Lao Tzu, op. cit., ch. 2.
14 Quoted in J. Needham, *Science and Civilization in China*, vol. II, p. 88.
15 Ibid., pp. 68–69.
16 Lao Tzu, op. cit., ch. 48.
17 Lao Tzu, op. cit., chs. 71, 72.
18 Chuang Tzu, op. cit., ch. 16.

Chapter 9: Zen

1 Chuang Tzu, trans. James Legge, ch. 22.

2 In A. W. Watts, *The Way of Zen*, p. 87.
3 In P. Reps, *Zen Flesh, Zen Bones*, p. 96.
4 In D. T. Suzuki, *Zen and Japanese Culture*, p. 16.
5 In P. Kapleau, *Three Pillars of Zen*, p. 49.
6 From the *Zenrin Kushu;* in A. W. Watts, op. cit., 134.

III THE PARALLELS

Chapter 10: The Unity of All Things

1 Ashvaghosha, *The Awakening of Faith*, p. 55.
2 Ibid., p. 93.
3 H. P. Stapp, "S-Matrix Interpretation of Quantum Theory," *Physical Review*, Vol. D3 (March 15th, 1971), pp. 1303–20.
4 Ibid., p. 1303.
5 N. Bohr, *Atomic Physics and the Description of Nature*, p. 57.
6 D. Bohm & B. Hiley, "On the Intuitive Understanding of Nonlocality as Implied by Quantum Theory," *Foundations of Physics*, Vol. 5 (1975), pp. 96, 102.
7 S. Aurobindo, *The Synthesis of Yoga*, p. 993.
8 Nagarjuna, quoted in T. R. V. Murti, *The Central Philosophy of Buddhism*, p. 138.
9 H. P. Stapp, op. cit., p. 1310.
10 W. Heisenberg, *Physics and Philosophy*, p. 107.
11 *Mundaka Upanishad*, 2.2.5.
12 W. Heisenberg, op. cit., p. 81.
13 Ibid., p. 58.
14 J. A. Wheeler, in J. Mehra (ed.), *The Physicist's Conception of Nature*, p. 244.
15 *Brihad-aranyaka Upanishad*, 4.5.15.
16 Chuang Tzu, trans. James Legge, ch. 6.
17 Lama Anagarika Govinda, *Foundations of Tibetan Mysticism*, p. 93.

Chapter 11: Beyond the World of Opposites

1 Lao Tzu, *Tao Te Ching, trans.* Ch'u Ta-Kao, ch. 1.
2 D. T. Suzuki, *The Essence of Buddhism*, p. 18.
3 Quoted in A. W. Watts, *The Way of Zen*, p. 117.
4 R. Wilhelm, *The I Ching or Book of Changes*, p. 297.
5 Lama Anagarika Govinda, *Foundations of Tibetan Mysticism*, p. 136.

6 V. F. Weisskopf, *Physics in the Twentieth Century—Selected Essays*, p. 30.
7 J. R. Oppenheimer, *Science and the Common Understanding*, pp. 42–43.
8 *Isa-Upanishad*, 5.
9 Ashvaghosha, *The Awakening of Faith*, p. 59.
10 Lama Anagarika Govinda, "Logic and Symbol in the Multidimensional Conception of the Universe," *Main Currents*, Vol. 25, p. 60.

Chapter 12: Space-time

1 In P. A. Schilpp (ed.), *Albert Einstein: Philosopher-Scientist*, p. 250.
2 *Madhyamika Karika Vrtti*, quoted in T. R. V. Murti, *The Central Philosophy of Buddhism*, p. 198.
3 J. Needham, *Science and Civilization in China*, vol. III, p. 458.
4 Ashvaghosha, *The Awakening of Faith*, p. 107.
5 M. Sachs, "Space-Time and Elementary Interactions in Relativity," *Physics Today*, Vol. 22 (February 1969), p. 53.
6 In A. Einstein et al., *The Principle of Relativity*, p. 75.
7 S. Aurobindo, *The Synthesis of Yoga*, p. 993.
8 D. T. Suzuki, Preface to B. L. Suzuki, *Mahayana Buddhism*, p. 33.
9 Chang Tzu, trans. James Legge, ch. 2.
10 Quoted in A. W. Watts, *The Way of Zen*, p. 201.
11 D. T. Suzuki, *On Indian Mahayana Buddhism*, pp. 148–49.
12 In P. A. Schilpp, op. cit., p. 114.
13 Lama Anagarika Govinda, *Foundations of Tibetan Mysticism*, p. 116.
14 Dogen Zenji, *Shobogenzo;* in J. Kennett, *Selling Water by the River*, p. 140.
15 Govinda, op. cit., p. 270.
16 S. Vivekananda, *Jnana Yoga*, p. 109.

Chapter 13: The Dynamic Universe

1 D. T. Suzuki, *The Essence of Buddhism*, p. 53.
2 Carlos Castaneda, *A Separate Reality*, p. 16.
3 S. Radhakrishnan, *Indian Philosophy*, p. 173.
4 *Brihad-aranyaka Upanishad*, 2.3.3.
5 *Bhagavad Gita*, 8.3.

6 Ibid., 3.24.
7 S. Radhakrishnan, op. cit., p. 367.
8 *Ts'ai-ken t'an*: quoted in T. Leggett, *A First Zen Reader*, p. 229, and in N. W. Ross, *Three Ways of Asian Wisdom*, p. 144.
9 A. C. B. Lovell, *The Individual and the Universe*, p. 93.
10 *Bhagavad Gita*, 9.7–10.
11 *Digha Nikaya*, ii, 198.
12 D. T. Suzuki, op. cit., p. 55.
13 J. Needham, *Science and Civilization in China*, vol. II, p. 478.

Chapter 14: Emptiness and Form

1 F. Hoyle, *Frontiers of Astronomy*, p. 304.
2 Quoted in M. Capek, *The Philosophical Impact of Contemporary Physics*, p. 319.
3 *Chandogya Upanishad*, 4.10.4.
4 *Kuan-tzu*, trans. W. A. Rickett, XIII, 36: a very large sociophilosophical work, traditionally attributed to the noted statesman Kuan Chung of the seventh century B.C. but most likely a composite work compiled around the third century B.C. and reflecting various philosophical schools.
5 *Chandogya Upanishad*, 3.14.1.
6 H. Weyl, *Philosophy of Mathematics and Natural Science*, p. 171.
7 Quoted in Fung Yu-lan, *A Short History of Chinese Philosophy*, p. 279.
8 Ibid., p. 280.
9 W. Thirring, "Urbausteine der Materie," *Almanach der Osterreichischen Akademie der Wissenschaften*, Vol. 118 (1968), p. 160.
10 J. Needham, *Science and Civilization in China*, vol. IV, pp. 8–9.
11 Lama Anagarika Govinda, *Foundations of Tibetan Mysticism*, p. 223.
12 *Prajna-paramita-hridaya Sutra,* in F. M. Muller (ed.), *Sacred Books of the East*, Vol. XLIX, "Buddhist Mahayana Sutras."
13 Quoted in J. Needham, op. cit., vol. II, p. 62.
14 Commentary to the hexagram *Yü*, in R. Wilhelm, *The I Ching or Book of Changes*, p. 68.

15 W. Thirring, op. cit., p. 159.
16 Quoted in J. Needham, op. cit., vol. IV, p. 33.

Chapter 15: The Cosmic Dance

1 K. W. Ford, *The World of Elementary Particles*, p. 209.
2 A. David-Neel, *Tibetan Journey*, pp. 186–87.
3 A. K. Coomaraswamy, *The Dance of Shiva*, p. 78.
4 H. Zimmer, *Myths and Symbols in Indian Art and Civilization*, p. 155.
5 A. K. Coomaraswamy, op. cit., p. 67.

Chapter 17: Patterns of Change

1 W. Heisenberg, *Physics and Philosophy*, p. 107.
2 G. F. Chew, "Impasse for the Elementary Particle Concept," *The Great Ideas Today* (William Benton, Chicago, 1974), p. 99.
3 Ashvaghosha, *The Awakening of Faith*, pp. 79, 86.
4 *Lankavatara Sutra*, in D. T. Suzuki, *Studies in the Lankavatara Sutra*, p. 242.
5 S. Radhakrishnan, *Indian Philosophy*, p. 369.
6 R. Wilhelm, *The I Ching or Book of Changes*, p. 315.
7 H. Wilhelm, *Change*, p. 19.
8 R. Wilhelm, op. cit., p. 348.
9 Ibid., p. 352.
10 R. Wilhelm, op. cit., p. 1.

Chapter 18: Interpenetration

1 G. F. Chew, " 'Bootstrap': A Scientific Idea," *Science*, Vol. 161 (May 23, 1968), pp. 762–65; "Hadron Bootstrap: Triumph or Frustration?," *Physics Today*, Vol. 23 (October 1970), pp. 23–28; "Impasse for the Elementary Particle Concept," *The Great Ideas Today* (William Benton, Chicago, 1974).
2 Quoted in J. Needham, *Science and Civilization in China*, vol. II, p. 538.
3 *G. F. Chew, "Bootstrap": A Scientic Idea?,* op. cit., pp. 762–63.
4 Lao Tzu, *Tao Te Ching*, trans. Ch'u Ta-Kao, ch. 25.
5 J. Needham, op. cit., vol. II, p. 582.
6 J. Needham, op. cit., vol. II, p. 484.
7 Ibid., pp. 558, 567.
8 Quoted in J. Needham, op. cit., vol. II, p. 566.

9 Ashvaghosha, *The Awakening of Faith*, p. 56.
10 In P. Reps, *Zen Flesh, Zen Bones*, p. 104.
11 Ibid., p. 119.
12 Ashvaghosha, op. cit., p. 104.
13 S. Aurobindo, *The Synthesis of Yoga*, p. 989.
14 D. T. Suzuki, *On Indian Mahayana Buddhism*, p. 150.
15 Ibid., pp. 183–84.
16 G. F. Chew, "Hadron Bootstrap: Triumph or Frustration?," op. cit., p. 7.
17 G. F. Chew, M. Gell-Mann and A. H. Rosenfeld, "Strongly Interacting Particles," *Scientific American*, Vol. 210 (February 1964), p. 93.
18 C. Eliot, *Japanese Buddhism*, pp. 109–10.
19 D. T. Suzuki, op. cit., p. 148.
20 D. T. Suzuki, *The Essence of Buddhism*, p. 52.
21 In P. P. Wiener, *Leibnitz—Selections*, p. 547.
22 In J. Needham, op. cit., vol. II, pp. 496 ff.
23 In P. P. Wiener, op. cit., p. 533.
24 Ibid., p. 161.
25 G. F. Chew, " 'Bootstrap': A Scientific Idea?," op. cit., p. 763.
26 E. P. Wigner, *Symmetries and Reflections—Scientific Essays*, p. 172.
27 G. F. Chew, " 'Bootstrap': A Scientific Idea?," op. cit., p. 765.
28 Lao Tzu, *Tao Te Ching*, trans. Ch'u Ta-Kao, ch. 81.

EPILOGUE

1 Lama Anagarika Govinda, *Foundations of Tibetan Mysticism*, p. 225.

BIBLIOGRAPHY

Alfven, H. *Worlds-Antiworlds*. San Francisco: W. H. Freeman, 1966.
Ashvaghosha. *The Awakening of Faith*. Transl. D. T. Suzuki. Chicago: Open Court, 1900.
Aurobindo, S. *The Synthesis of Yoga*. Pondicherry, India: Aurobindo Ashram Press, 1957.
——. *On Yoga II*. Pondicherry, India: Aurobindo Ashram Press, 1958.
Bohm, D. and Hiley, B. *On the Intuitive Understanding of Nonlocality as Implied by Quantum Theory*. Foundations of Physics, Vol. 5, 1975, pp. 93–109.
Bohr. N. *Atomic Physics and Human Knowledge*. New York: John Wiley & Sons, 1958.
——. *Atomic Physics and the Description of Nature*. Cambridge, Eng.: Cambridge University Press, 1934.
Capek, M. *The Philosophical Impact of Contemporary Physics*. Princeton, N.J.: D. Van Nostrand, 1961.
Castaneda, C. *The Teachings of Don Juan*. New York: Ballantine Books, 1968.
——. *A Separate Reality*. New York: Simon and Schuster, 1971.
——. *Journey to Ixtlan*. New York: Simon and Schuster, 1972.
——. *Tales of Power*. New York: Simon and Schuster, 1974.
Chew, G. F. " 'Bootstrap': A Scientific Idea?," *Science*, Vol. 161, pp. 762–65, May 23, 1968.
——. "Hadron Bootstrap: Triumph or Frustration?," *Physics Today*, Vol. 23, pp. 23–28, October 1970.
——. "Impasse for the Elementary Particle Concept," The Great Ideas Today, 1974, Chicago, Ill.: *Encyclopaedia Britannica*, 1974.
Chew, G. F., Gell-Mann, M. and Rosenfeld, A. H. "Strongly Interacting Particles," *Scientific American*, Vol. 210, pp. 74–83, February 1964.
Chuang Tzu. Transl. James Legge, arranged by Clae Waltham, New York: Ace Books, 1971.
Chuang Tzu. *Inner Chapters*. Transl. Gia-Fu Feng and Jane English, New York: Vintage Books, 1974.
Coomaraswamy, A. K. *Hinduism and Buddhism*. Philosophical Library, New York, 1943.

——. *The Dance of Shiva*. New York: The Noonday Press, 1969.

Crosland, M. P. (ed.). *The Science of Matter*. History of Science Readings, Baltimore, Md.: Penguin Books, 1971.

David-Neel, A. *Tibetan Journey*. London: John Lane, 1936.

Einstein, A., *Essays in Science*. New York: Philosophical Library, 1934.

——. *Out of My Later Years*, New York: Philosophical Library, 1950.

Einstein, A. et al., *The Principle of Relativity*. New York: Dover, 1923.

Eliot, C. *Japanese Buddhism*. New York: Barnes & Noble, 1969.

Feynman, R. P., Leighton, R. B. and Sands, M. *The Feynman Lectures on Physics*. Reading, Mass.: Addison-Wesley, 1966.

Ford, K. W. *The World of Elementary Particles*. New York: Blaisdell, 1965.

Fung, Yu-lan. *A Short History of Chinese Philosophy*. New York: Macmillan, 1958.

Gale, G. "Chew's Monadology." *Journal of History of Ideas*, Vol. 35, pp. 339–48, April–June 1974.

Govinda, L. A. *Foundations of Tibetan Mysticism*. New York: Samuel Weiser, 1974.

——. "Logic and Symbol in the Multidimensional Conception of the Universe," *Main Currents*, Vol. 25, pp. 59–62, 1969.

Guthrie, W. K. C. *A History of Greek Philosophy*. Cambridge, Eng.: Cambridge University Press, 1969.

Heisenberg, W. *Physics and Philosophy*. New York: Harper Torchbooks, 1958.

——. *Physics and Beyond*, New York, Harper & Row, 1971.

Herrigel, E. *Zen in the Art of Archery*. New York: Vintage Books, 1971.

Hoyle, F. *The Nature of the Universe*. New York: Harper, 1960.

——. *Frontiers of Astronomy*. New York: Harper, 1955.

Hume, R. E. *The Thirteen Principal Upanishads*. New York: Oxford University Press, 1934.

James, W. *The Varieties of Religious Experience*. New York: Longmans, Green & Co., 1935.
Jeans, J. *The Growth of Physical Science*. Cambridge, Eng.: Cambridge University Press, 1951.
Kapleau, P. *Three Pillars of Zen*. Boston: Beacon Press, 1967.
Kennett, J. Selling Water by the River. New York: Vintage Books, 1972.
Keynes, G. (ed.). *Blake—Complete Writings*. New York: Oxford University Press, 1969.
Kirk, G. S. *Heraclitus—The Cosmic Fragments*. Cambridge, Eng.: Cambridge University Press, 1970.
Korzybski, A. *Science and Sanity*. Lakeville, Conn.: The International Non-Aristotelian Library, 1958.
Krishnamurti, J. *Freedom from the Known*. New York: Harper & Row, 1969.
Kuan Tzu. Transl. W. A. Rickett. Hong Kong University Press, 1965.
Lao Tzu. *Tao Te Ching*, transl. Ch'u Ta-Kao. New York: Samuel Weiser, 1973.
Lao Tzu. *Tao Te Ching*, transl. Gia-fu Feng and Jane English. New York: Vintage Books, 1972.
Leggett, T. *A First Zen Reader*. Rutland, Vermont: C. E. Tuttle, 1972.
Lovell, A. C. B. *The Individual and the Universe*. New York: Harper, 1959.
——. *Our Present Knowledge of the Universe*. Cambridge, Mass.: Harvard University Press, 1967.
Maharishi Mahesh Yogi. *Bhagavad Gita*. Chapters 1–6, transl. and commentary, Baltimore, Md.: Penguin Books, 1973.
Mascaro, J. *The Bhagavad Gita*. Baltimore, Md.: Penguin Books, 1970.
——. *The Dhammapada*. Baltimore, Md.: Penguin Books, 1973.
Mehra, J. (ed.). *The Physicist's Conception of Nature*. D. Reidel, Dordrecht-Holland, 1973.
Miura, I. and Fuller-Sasaki, R. *The Zen Koan*. New York: Harcourt Brace & World, 1965.
Muller, F. M. (ed.). *Sacred Books of the East*. Vol. XLIX,

Buddhist Mahayana Sutras, New York: Oxford University Press.

Murti, T. R. V. *The Central Philosophy of Buddhism.* London: Allen & Unwin, 1955.

Needham, J. *Science and Civilization in China.* Cambridge, Eng.: Cambridge University Press, 1956.

Oppenheimer, J. R. *Science and the Common Understanding.* New York: Oxford University Press, 1954.

Radhakrishnan, S. *Indian Philosophy.* New York: Macmillan, 1958.

Reps, P. *Zen Flesh, Zen Bones.* New York: Anchor Books.

Ross, N. W. *Three Ways of Asian Wisdom.* New York: Simon & Schuster, 1966.

Russell, B. *History of Western Philosophy.* New York: Simon & Schuster, 1945.

Sachs, M. "Space-Time and Elementary Interactions in Relativity," *Physics Today*, Vol. 22, pp. 51–60, February 1969.

Sciama, D. W. *The Unity of the Universe.* London: Faber and Faber, 1959.

Schilpp, P. A. (ed.). *Albert Einstein: Philosopher-Scientist,* Evanston, Ill.: The Library of Living Philosophers, 1949.

Stace, W. T. *The Teachings of the Mystics.* New York: New American Library, 1960.

Stapp, H. P. "S-Matrix Interpretation of Quantum Theory" *Physical Review*, Vol. D3, pp. 1303–20, March 15, 1971.

Suzuki, D. T. *The Essence of Buddhism.* Kyoto, Japan: Hozokan, 1968.

——. *Outlines of Mahayana Buddhism.* New York: Schocken Books, 1963.

——. *On Indian Mahayana Buddhism.* E. Conze (ed.). New York: Harper & Row, 1968.

——. *Zen and Japanese Culture.* New York: Bollingen Series, 1959.

——. *Studies in the Lankavatara Sutra.* London: Routledge & Kegan Paul, 1952.

——. Preface to B. L. Suzuki, *Mahayana Buddhism.* London: Allen & Unwin, 1959.

Thirring, W. "Urbausteine der Materie." *Almanach der Osterreichischen Akademie der Wissenschaften*, Vol. 118, pp. 153–62. Vienna, Austria, 1968.

Vivekananda, S. *Jnana Yoga*. New York: Ramakrishna-Vivekananda Center, 1972.

Watts, A. W. *The Way of Zen*. New York: Vintage Books, 1957.

Weisskopf, V. F. *Physics in the Twentieth Century*. Selected Essays, Cambridge, Mass.: M.I.T. Press, 1972.

Weyl, H. *Philosophy of Mathematics and Natural Science*. Princeton, N.J.: Princeton University Press, 1949.

Whitehead, A. N. *The Interpretation of Science*. Selected Essays, A. H. Johnson (ed.). Indianapolis, N.Y.: Bobbs-Merrill, 1961.

Wiener, P. P. *Leibnitz—Selections,* New York: Ch. Scribner's Sons, 1951.

Wigner, E. P. *Symmetries and Reflections*. Scientific Essays, Cambridge, Mass.: M.I.T. Press, 1970.

Wilhelm, H. *Change—Eight Lectures on the I Ching*. New York: Harper Torchbooks, 1964.

Wilhelm, R. *The I Ching or Book of Changes*. Princeton, N.J.: Princeton University Press, 1967.

——. *The Secret of the Golden Flower*. London: Routledge & Kegan Paul, 1972.

Woodward, F. L. (transl. and ed.). *Some Sayings of the Buddha*. New York: Oxford University Press, 1973.

Zimmer, H. *Myths and Symbols in Indian Art and Civilization*. Princeton, N.J.: Princeton University Press, 1972.

INDEX